DR. BOUALAMALLAH DJAMAL

Design and optimization of a solar water heater

DR. BOUALAMALLAH DJAMAL

Design and optimization of a solar water heater

Design and Optimization

ScienciaScripts

Imprint

Cover image: www.ingimage.com

This book is a translation from the original published under ISBN 978-620-6-70467-6.

Publisher:
Sciencia Scripts
is a trademark of
Dodo Books Indian Ocean Ltd. and OmniScriptum S.R.L publishing group

120 High Road, East Finchley, London, N2 9ED, United Kingdom
Str. Armeneasca 28/1, office 1, Chisinau MD-2012, Republic of Moldova, Europe
Printed at: see last page
ISBN: 978-620-8-07017-5

Contents

Acknowledgements

First of all, thanks to the great GOD who gave me the will and the courage to complete this work, and through this modest work, I would like to express my respectful gratitude to my two promoters "DR. A. Merdji and DR. DELLA NOUREDDINE", for their encouragement and precious advice, and for all the comforts and ease they gave us during the study and realisation of this project. Our warmest thanks also go to the members of the jury for agreeing to examine and evaluate our work. My thanks and esteem go to all the professors and teachers in the mechanical engineering department. Without forgetting of course to deeply thank all those who have contributed in any way to the realization of this work.

BOUALAMALLAH DJAMAL

General introduction

The use of solar thermal energy is making a comeback, thanks to its huge impact on reducing CO_2 emissions and to high-performance installations. For collective installations, the implementation of a solar performance guarantee is a must. To harness or store this solar energy, it has to be converted into another form of energy, which is why solar collectors are used.

The issue of energy has dominated the world's economic and political scenes over the last forty years. Never has an issue raised so many stakes and created so much controversy. The forced recourse to the domination of energy sources such as oil and gas fields is at the root of several regional conflicts. The energy crisis of 1973, triggered by the October war, acted as a detonator, shaking the world out of its lethargy and prompting it to think about alternatives to existing conventional resources, which, despite their preponderance, are not inexhaustible. Most of these resources are hydrocarbons, and there is a risk that they will run out in the very near future if they are not used rationally.

Solar energy is one of the most readily exploitable new forms of energy, and in recent years it has enjoyed a boom in terms of the diversity of its applications and the interest it has aroused around the world. However, the high cost of solar energy compared with conventional energy sources is a handicap to the long-awaited expansion in the use of solar energy.

Optimising solar energy systems is one of the solutions recommended to help reverse the current trend and see solar energy applications become more widespread around the world.

The simplest and most immediate application of solar energy is the production of hot water for domestic use. It is also one of the oldest, since several solar water heating systems have been designed around the world from the early twentieth century to the present day, each more efficient than the last.

A solar heating system generally consists of three parts: collection, storage and distribution. Collection is the main part of solar conversion. It consists of the solar collector. It is the collector that converts the sun's energy into heat, which it transmits to the heat transfer fluid contained in its absorber. Given the very important role played by the collector or solar collector in the process of converting solar energy into thermal energy, a number of studies have focused on the flat-plate solar collector, with the main aim of improving its efficiency, which is the most significant performance factor. The efficiency of the flat plate solar collector varies according to the geometry of the collector, its internal parameters and external parameters such as sunshine, ambient temperature, etc.

The aim of our work is to compare two types of collector in order to optimise the glazed flat-plate collectors and vacuum tube collectors, and to find the optimum angle during the year for good operation.

To this end, a system of equations governing the thermal behaviour of the flat plate solar water collector in the transient regime is established. The results obtained are represented graphically, commented on and interpreted, and a general conclusion and advantages are given towards the end. Four chapters, an introduction and a general conclusion form the backbone of this work. In the first chapter, we present the astronomical notions necessary for any study of the solar system. In the second chapter, a general introduction to solar water heaters and their operating principles is given, followed by a theoretical study of the flat-plate solar collector, its various components, its classifications, the different parameters that influence its efficiency, its operating mode and the equations that govern its behaviour in transient conditions. This is the subject of the third chapter, followed by a simulation using TRNSYS software, which enables me to simulate a solar water heater according to the weather at the Adrar station and to interpret the results.

CHAPTER 1

THE DEPOSIT SOLAR

Introduction :

Energy comes from nature in two forms:

- Conventional or non-renewable energy sources are fossil fuels, the best known of which are oil, coal, gas and uranium.
- Renewable or non-conventional energies, the most important of which are: solar, wind, geothermal and biomass.

They originate from inexhaustible energy sources thanks to natural cycles such as solar radiation, wind, the earth's internal heat flow and the carbon cycle in the biosphere.

Solar energy is the most dominant of all renewable energies, and one of the easiest to harness. Like most soft energies, it allows users to meet some of their needs without any intermediary.

Knowing the position of the sun in the sky at any time and in any place is necessary for studying the energy intercepted. The times of sunrise and sunset and the path of the sun across the sky over the course of a day can be used to assess certain variables such as the maximum duration of insolation and global irradiation.

In this section we will define certain solar parameters, namely :

- Astronomical quantities.
- Geographical quantities.
- Solar radiation outside the atmosphere.
- Direct, diffuse and global radiation.

1.1 The sun

The sun is the main source of all forms of energy on earth. This is as true for conventional fossil fuels, such as hydrocarbons, which are the result of photosynthesis, as it is for non-conventional renewable energies, such as solar, wind, biomass and geothermal energy.

The Sun is gaseous, spherical and 14x105 km in diameter, with a mass of around 2x1030 kg. It is made up mainly of 80% hydrogen and 19% helium, with the remaining 1% a mixture of more than 100 elements [1], [2].

It is located at a distance of about 150 million km from the Earth. Its total luminosity, i.e. the power it emits in the form of photons, is roughly equal to 4x1026w. Only part of this is intercepted by the Earth, at around 1.7x1017w. It reaches us mainly in the form of electromagnetic waves; 30% of this power is reflected back into space, 47% is absorbed and 23% is used as a source of energy for the evaporation-precipitation cycle in the atmosphere [1], [2].

The main characteristics of the sun are listed in the table below [3] :

Table I.1: Main characteristics of the sun [3].

Diameter (km)	14x105
Weight (kg)	2x1030
Surface area (km)^2	6.09x1012
Volume (km3)	18 1.41x10
Average density (kg/m3)	1408
Speed (km/s)	217
Distance from the centre of the Milky Way (km)	47 2.5x10

The sun is not a homogeneous sphere, and three main regions can be distinguished (fig.1.1) [1], [2], [3] :

a/ The interior, which contains 40% of the Sun's mass, is where energy is created by thermonuclear reaction. This region extends over a thickness of 25x104 km. This layer is divided into three zones: the core, the radiative zone and the convective zone. The radiation emitted in this part is totally absorbed by the upper layers. The temperature reaches several million degrees, and the pressure a billion atmospheres.

b/ The photosphere is a very thin, opaque layer, about 300 km thick, which is responsible for almost all the radiation we receive, the visible part of the Sun. The order of magnitude of the temperature is only a few million degrees, decreasing very rapidly with the thickness of the layer until it reaches a so-called surface temperature of around 4500 .

c/ The solar chromosphere and corona are regions of low density where matter is very dilute. This layer is characterised by very low emitted radiation, although the temperature is very high (one million degrees),

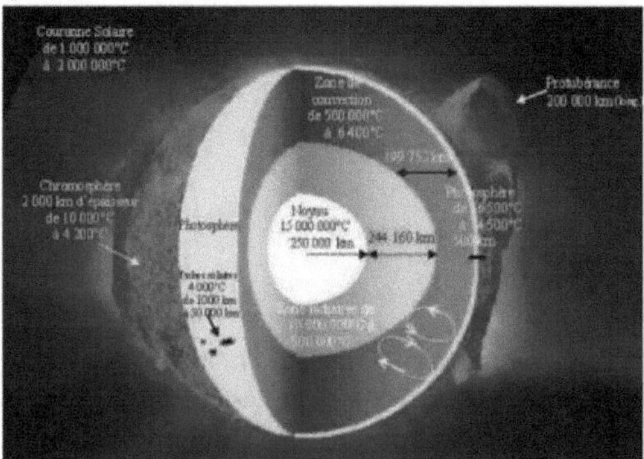

Figure I.1 Structure of the sun

1.2 The movement of the earth

In its movement around the Sun, the Earth describes an ellipse with the Sun as one of its foci. The complete revolution takes place in a period of 365.25 days. The plane of this ellipse is called the ecliptic [1].

It is at the winter solstice (21 December) that the Earth is closest to the Sun: 147 million km. On 22 June, the distance from the Earth to the Sun is 152 million km. This is the day when

the Earth is furthest away, the summer solstice. 21 March and 21 September are called the vernal and autumnal equinoxes respectively. At the equinoxes, day and night are equal [1].
In addition to its rotation around the sun, the earth also turns on itself around an axis called the pole axis. This rotation takes place in one day. The plane perpendicular to the pole axis and passing through the centre of the Earth is called the equator. The axis of the poles is not perpendicular to the ecliptic; they form an angle between them called an inclination equal to 23°27' [1].

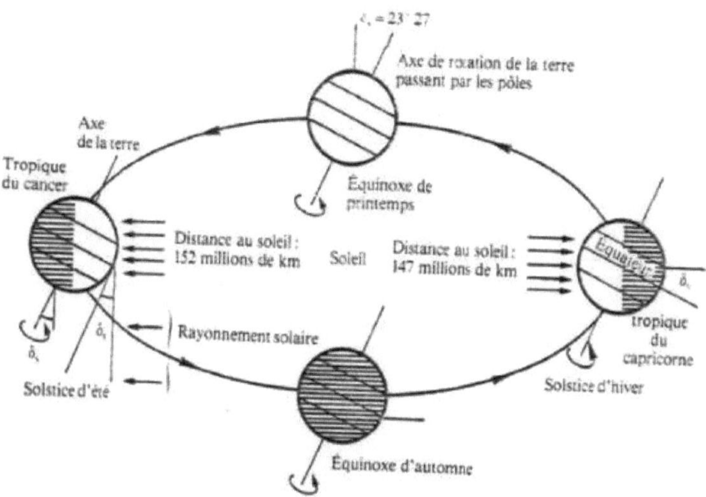

1.3 Geographical and astronomical dimensions

I.3.1 Dimensions for locating the sun

The position of the sun in the sky is not fixed; it changes throughout the day and the season. This change in position is caused by the earth's rotation on itself (around its axis) and its movement around the sun (in its orbit) [1], [4].
To determine this position, it is usual to use two reference marks: the equatorial or hourly mark and the horizontal or azimuthal mark.

- **.3.1.1 Equatorial coordinate marker**

In this reference frame, the position of the sun in the sky is determined by two quantities [1], [4]:

- **Declination (δ):** This is the angle between the direction of the sun and the earth and the plane of the earth's equator. It is zero at the equinoxes and maximum at the solstices, varying from -23.27° at the winter solstice to +23.27° at the summer solstice.

As a first approximation, it can be evaluated by the following relationship:

$$\delta = 23{,}27 \sin [360 /3655(j + 284)] \qquad (I.1)$$

is expressed in degrees.
j is the number of the day of the year from the first of January.

- **Hour angle (ω):** This is the angle between the prime meridian passing through South and the projection of the sun onto the equatorial plane, it measures the sun's path across the sky. It is given by the following relationship :

$$\omega = 15\ (TSV - 12) \qquad (I.2)$$

TSV: true solar time ;

It is 0° at solar noon, then each hour corresponds to a variation of 15°, because the period of the earth's rotation on itself is equal to 24 hours. Counted negatively in the morning when the sun is east and positively in the evening.

True solar time is equal to legal time corrected by an offset due to the difference between the longitude of the location and the reference longitude.

$$ET = 9{,}87 \sin (2JD) - 7{,}35 \cos (JD) - 1{,}5\ (JD) \qquad (I.4)$$

With $$JD = (J - 81) \times (360/365) \qquad (I.5)$$

TU: Universal Standard Time (min)

LSt: standard meridian of the location (°)

Lg: local meridian of the location (°)

ET: correction of the equation of time, given by the following relationship :

$$ET = 9.87 \sin (2JD) - 7.35 \cos (JD) - 1.5\ (JD) \qquad (I.4)$$

Where $JD = (J - 81) \times (360/365)$ (I.5)

And **D**: number of days from the first of January.

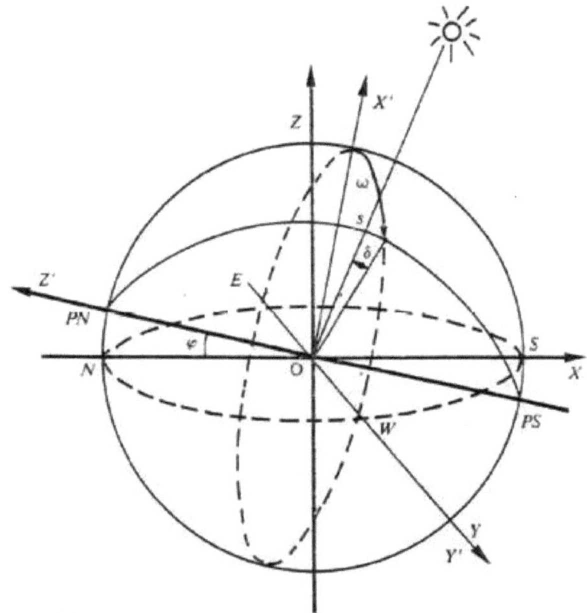

Figure I.3 Time coordinate marker

I.3.1.2 Horizontal coordinate marker

The horizontal reference point is formed by the plane of the astronomical horizon and the vertical of the location [1].

In this reference frame, the coordinates of the sun are :

- **The height of the sun (h):** This is the angle formed by the direction of the sun and its projection on the horizontal plane.
It is given by the following relationship :

$$\mathbf{Sin(h) = sin(\Phi).sin(\delta) + cos(\Phi).cos(\delta)} \quad (I.6)$$

Φ: latitude of the location.
- **Azimuth of the sun (*a*):** This is the angle between the projection of the sun's direction on the horizontal plane and the south. The azimuth is counted positively towards the west and negatively towards the east.
It is given by the following relationship :

$$\mathbf{Sin\ (a) = (cos(\delta).sin(\omega))/cos(h)} \quad (I.7)$$

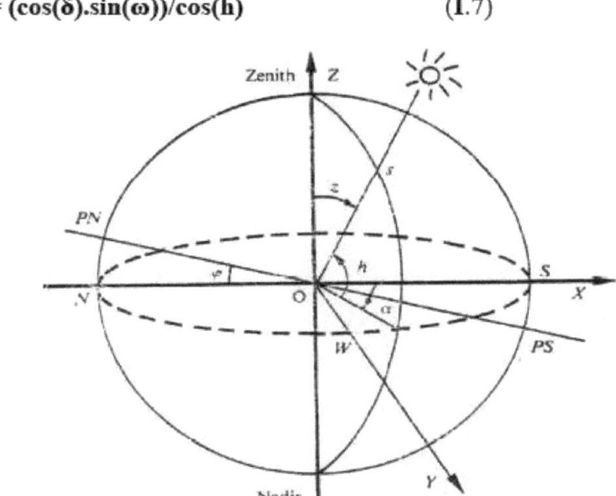

Figure I.4 Azimuthal coordinate marker

1.3.2 Quantities for locating a site on the earth's surface

Any point on the globe can be defined by the following coordinates [1], [4]:
- **Latitude (Φ):** corresponds to the angle between the radius joining the centre of the earth at this location and the equatorial plane. It varies from -90° to +90° and is positive towards the north.
- **Longitude (The):** represents the angle between the meridian plane passing through this location and the original meridian plane (Green winch).

The altitude is the vertical distance between this point and a theoretical reference surface (surface
of the sea).

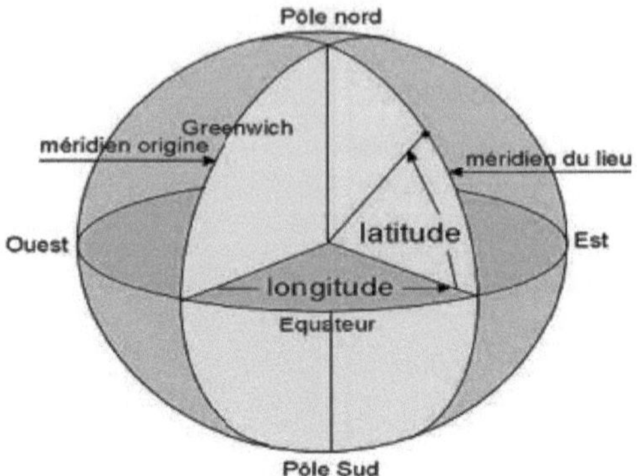

Figure I.5 Locating a site on the earth's surface

1.1.3 Plan orientation

The orientation of any plane is defined by two angles (a, γ) [1], [4]:

a: azimuth of the plane, the angle between the projection of the normal on the horizontal plane and the direction of south.

γ: height of the plane, this is the angle made by the normal of the plane and its projection onto the horizontal plane. The inclination β of the plane with respect to the horizontal plane is given by :

$$\beta = 90 - \gamma \qquad \text{(I.8)}$$

1.1.4 Angle of incidence on a plane

The angle of incidence *i* is the angle between the direction of the incident solar rays and the normal to the plane of the receiving surface. It is given by the following relationship [1], [4]:

$$\mathrm{Cos}(i) = \cos(\alpha - a) \,.\, \cos(\gamma) \,.\cos(h) + \sin(\gamma) \,.\, \sin(h) \qquad \text{(I.9)}$$

1.4 Solar radiation

The electromagnetic radiation emitted by the Sun is the external manifestation of the nuclear interactions occurring at the Sun's core and of all the secondary interactions they generate in its envelope. It accounts for almost all the energy expelled by the Sun.

Solar radiation is characterised by various features, the most important of which is the solar constant, which is a fundamental datum that is independent of meteorological conditions [5].

The second characteristic of solar radiation is its spectral distribution, which is roughly that of a black body at 5800°k. Solar radiation is a superposition of waves with lengths ranging from 0.25 micrometres to 4 micrometres [1], [5].

Figure I.6 shows the solar spectrum [6].

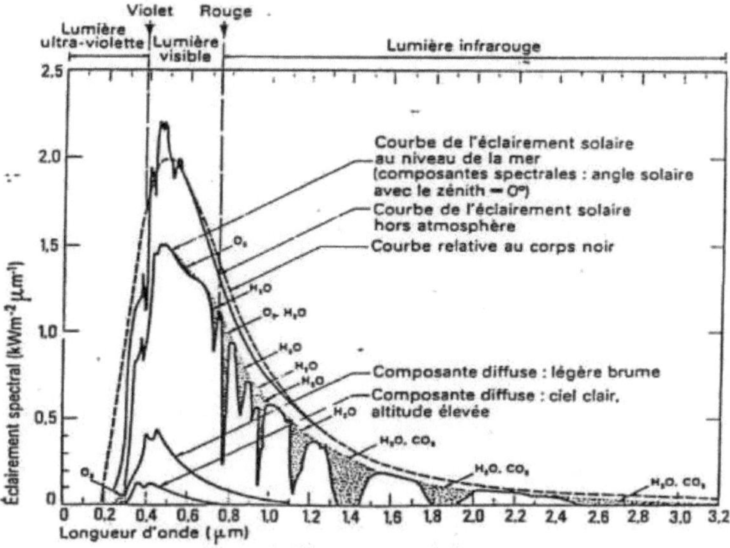

Figure 1.6 Le spectre solaire

1.4.1 Solar radiation outside the atmosphere

Solar radiation outside the atmosphere depends solely on astronomical parameters and is characterised by a fundamental datum known as the solar constant [1], [5].

The solar constant :

The solar constant E_0 is the energy flux received by a unit surface, normal to the sun's rays, located outside the atmosphere at an average distance from the earth to the sun.

Numerous experiments have been carried out to measure the solar constant. In our case, we will adopt a value of 1353 w/m2 (± 1.5%). This flux, known as the solar constant, varies slightly over the course of the year, depending on variations in the distance from the earth to the sun.

As a first approximation, the value of E can be calculated as a function of the number of the day in year d using the following relationship:

$$E = E_0 \, [1 + 0{,}033 \cos(0{,}984\, j)] \qquad (1.10)$$

The annual trend is shown in Figure 1.7 below:

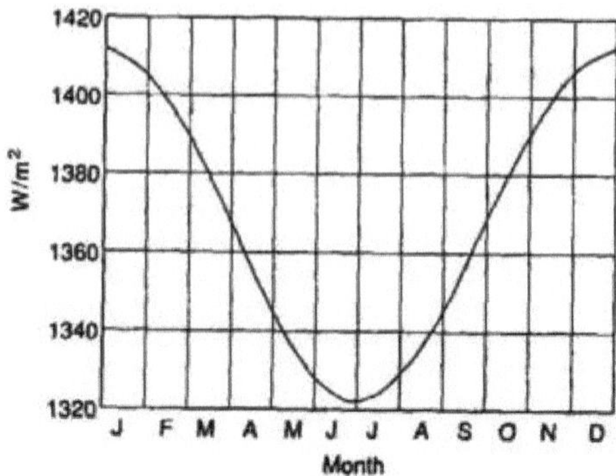

Figure I.7 Annual change in the solar constant

We can see that the maximum is obtained in January with a value of 1413w/m, the minimum at the beginning of July with a value of 1320w/m2.

1.4.2 Solar radiation received at ground level

The Earth's atmosphere greatly disrupts the flow of photons from the sun through a variety of processes. Therefore, after passing through the atmosphere, solar radiation can be considered as the sum of two components [1], [5] :

- **Direct radiation** is that which passes through the atmosphere unchanged. It comes from the sun's disc only, to the exclusion of any radiation that is scattered, reflected or refracted by the atmosphere.
- **Scattered radiation** is the part of solar radiation that comes from the entire celestial vault, with the exception of the solar disc, and is scattered by solid or liquid particles suspended in the atmosphere. It has no preferred direction.

Global radiation is the radiation received on a horizontal surface from the sun and the entire sky. It is the sum of direct and diffuse radiation. Figure 1.8 illustrates the different components of solar radiation at ground level.

The three quantities, direct radiation I, diffuse radiation D, and global radiation G, are linked by the following relationship:

$$\mathbf{G = I \cdot \sin(h) + D} \qquad (1.11)$$

Where h is the height of the sun.

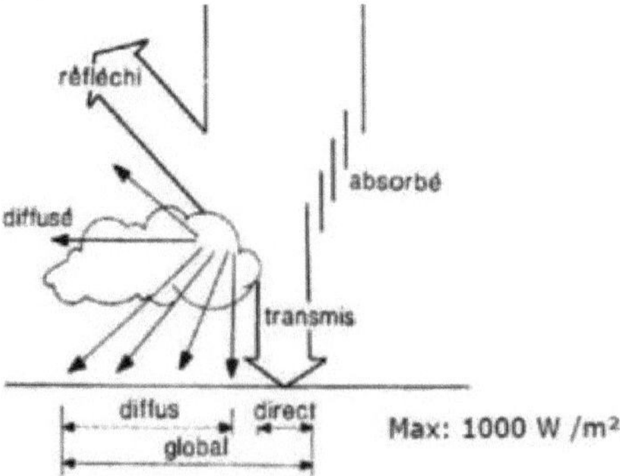

Figure I.8 Direct, diffuse and global radiation

1.5 . Solar energy in Algeria

The solar resource is a set of data describing changes in available solar radiation over a given period. It is used to simulate the operation of a solar energy system and to design it as accurately as possible, taking into account the demand to be met [4].

Because of its geographical location, Algeria has an enormous solar resource, as shown in Figure I.9 :

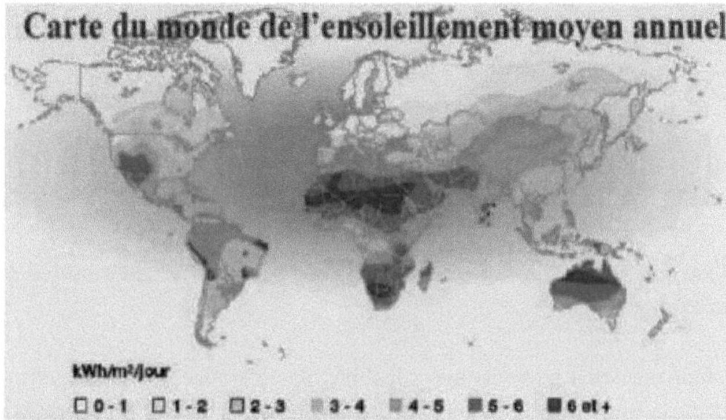

Figure I.9 World map of mean annual sunshine

Following a satellite evaluation, the German Space Agency (ASA) has concluded that Algeria has the greatest solar potential in the whole Mediterranean basin: 169,000 TWh/year for solar thermal energy and 13.9 TWh/year for solar photovoltaic energy. Algeria's solar potential is the equivalent of 10 large deposits of natural gas that would have been discovered at Hassi R'Mel. Table 1.2 shows the distribution of Algeria's solar potential by climatic region, according to the amount of sunshine received each year [7]:

Table I.2 Algerian insolation by climatic region [7].

Regions	Coastal regions	High plateaux	Sahara
Surface area (%)	4	10	86
Average sunshine duration (h/year)	2650	3000	3500
Average energy received (kWh/m2/year)	1700	1900	26500

The duration of sunshine in the Algerian Sahara is around 3,500 hours per year, the highest in the world. It is always more than 8 hours per day and can be as much as 12 hours per day during the summer, with the exception of the extreme south, where it drops to 6 hours per day during the summer.

The Adrar region is particularly sunny and has the greatest potential in the whole of Algeria (Figure I.10) [7].

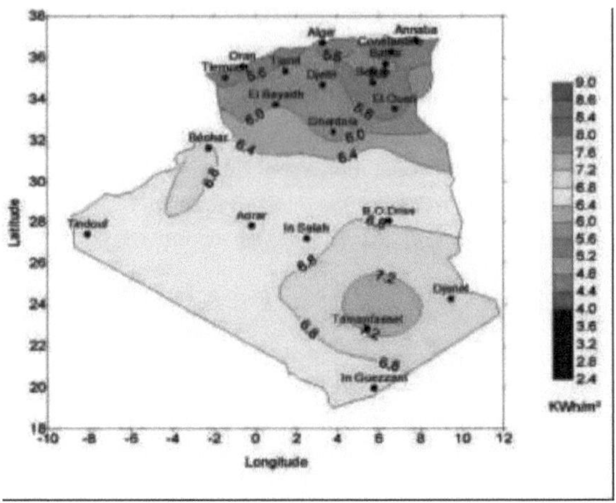

Figure I.10 Average annual global solar irradiance received on an inclined plane at the latitude of the location

Conclusion:

This chapter has provided a summary of some basic concepts relating to solar energy. Knowledge of these fundamental concepts, and in particular global radiation at ground level, will be of great use to us in exploiting solar energy using solar collectors and concentrators.

Solar energy is available over the entire surface of our planet, which receives more than 15,000 times the energy that mankind consumes. This energy can be harnessed in three ways: thermal energy, thermodynamic energy and photovoltaic energy.

Algeria has a large as yet unexploited solar resource. This form of energy offers many advantages in terms of thermal conversion, mainly for heating and electricity generation. It is an available, economical, non-polluting form of energy that requires little maintenance.

Solar measurements are mainly ground-based measurements of direct, diffuse and global radiation. Other parameters can also be measured: sunshine duration and hourly time.

Solar measurements are carried out by devices such as heliographs, pyrheliometers, which measure the incident ray, and pyranometers.

CHAPTER 2

GENERAL ON HEAT SOLAR WATER

11.1Introduction

Solar energy is the energy of the future from the point of view of energy efficiency and environmental protection. To harness or store this solar energy, it has to be converted into another form of energy, which is why solar collectors are used.

11.2Solar collectors

Depending on the energy conversion, there are two categories of solar collectors:

- Thermal solar collectors.
- Photovoltaic solar collectors.

Our work will focus solely on the solar thermal collector.

- **I.2.1 Thermal solar collectors**

These are collectors that convert solar energy into thermal energy, used for space heating and low-temperature domestic hot water production. There are two categories of solar thermal collectors:

- Liquid flow sensors.
- Air sensors.

- **.2.1.1 Liquid flow sensors**

These are collectors where the heat transfer fluid circulating through a solar circuit is a liquid (water, oil, thermal fluid, antifreeze).

The most common liquid solar collectors are :

- flat-plate sensors.
- concentration sensors.

a) Planar sensor (or isolator) [8]

There are three types of flat-plate solar collectors:

- Flat glass sensors.
- Non-glazed flat-plate sensors.
- High-performance flat-plate sensors.

a.1) Flat glass collectors

This is a very simple element, comprising a metal absorber that converts solar radiation into heat and transmits this heat to a heat-transfer liquid. The absorber is mounted in an insulated casing covered with a highly transparent glass or synthetic sheet. The absorber has a black

layer, often selective, which effectively absorbs solar radiation and reduces losses due to the three modes of heat transfer. [8]
For temperatures between 35° and 90°c, it is necessary to use collectors with glazing.
In this case, the absorber is made of metal (copper or aluminium), in a box insulated at the back and glazed at the front.
The function of glazing is to trap radiation by creating a greenhouse effect.
This type of collector is generally used to produce DHW (domestic hot water).
The following diagram shows the structure of a flat-glass collector. Figure (I.1).

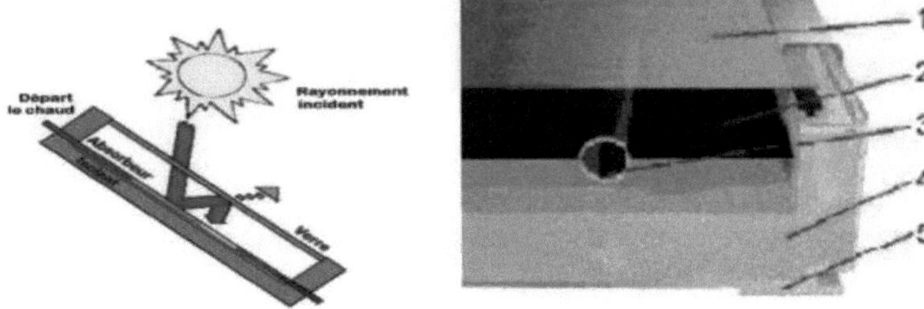

Figure II.1: Cross-section of a glazed flat-plate solar collector

1. One or more transparent cutlery sets.
2. An absorbent plate.
3. A hydraulic circuit for evacuating the fluid.
4. Thermal insulation.
5. A bin to hold it all.

a.2) Unglazed flat plate collectors

This collector is the simplest imaginable. Its usual application is heating outdoor swimming pools, but it cannot be used to produce DHW, except in hot countries.
metal coated with a selective layer, consists of a network of black tubes joined together. [8]

a.3) High performance planar sensors

a.3.1) Selective planar sensors

Some absorbers have a selective coating whose property is to emit only a small portion of the energy absorbed (7 to 20%). For most sensors, this type of coating is based on nickel and chromium.
The selective absorber improves the efficiency of the collector.
This feature is of particular interest in cold climates and for applications requiring high temperatures (water).figure(**II**.2)

Reveternentselectif

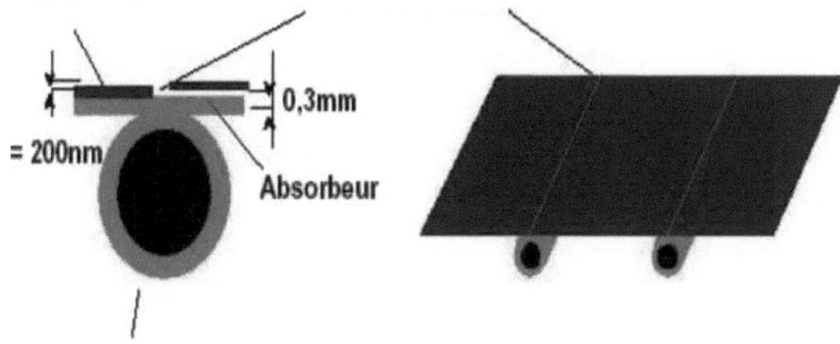

Pipe

Figure II.2: Absorber and selective coating

a.3.2) Flat vacuum sensors

Vacuum collectors can reach higher temperatures (up to 120°C). They consist of glass tubes containing a selective absorber.

The vacuum in the tubes greatly reduces heat loss from the collector. Another advantage of this collector is that it can be positioned at any angle, making it easy to integrate.

These collectors are well suited to the production of DHW (hot water) in the mountains or in northern countries, given their good efficiency at low outdoor temperatures. See Figure (II.3). [8]

There are three types of vacuum tubes:

- ❖ Direct-flow vacuum tubes.
- ❖ Vacuum caladuc tubes.
- ❖ Sydney" vacuum tubes.

Figure II.3: Vacuum sensor

b) Concentration sensors [8]

b.1) Parabolic trough concentrating collectors

Concentrating collectors work by tracking the sun. Depending on whether the light is concentrated on a point (tracking the sun along two axes) or on a line (tracking the sun along one axis), the temperatures reached are higher or lower. Parabolic trough collectors concentrate the light on a linear absorber, tracking the sun in a single direction. As a result, the concentration factor is not very high, and neither are the temperatures reached. These systems can reach temperatures of 200 to 400°C, for power ratings of several hundred kW.

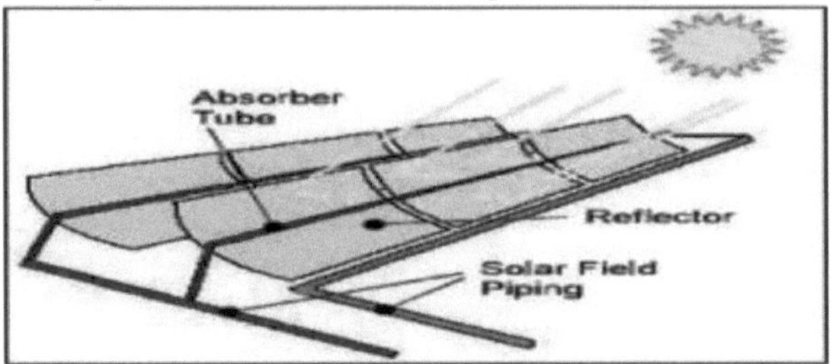

Figure II.4: Diagram of a parabolic trough transducer.

The first major industrial power plants of this type were developed in the United States by the Luz Solar company. There are two variants, depending on the moving part:

- Or the mirror
- Either the absorber

Figure II.5: Diagram showing the moving parts of a parabolic trough transducer

b.2) Dish-type concentrating collectors

Dish-type concentrating collectors concentrate the light on a point absorber, with the sun tracking in two directions. As a result, the concentration factor is higher, as are the temperatures reached. These systems can reach temperatures of 400 to 800°C, for power ratings of several tens of kW.

Figure II.6: Diagram of a dishes-type concentration sensor.

b.3) Tower power plants

Installations of this type concentrate the light on a point absorber, using mirrors (called "heliostats") that follow the sun in two directions, on a boiler located at the top of a tower. As a result, the concentration factor is greater, as are the temperatures reached. Given the large number of mirrors, power output can reach several MW. A plant of this type (Thémis) was tested in the 1980s, and is due to come back into service shortly. These systems can reach temperatures of between 400 and 800°C, for power outputs of several MW.

Figure II.7 : Tower solar power plant

II.2.1.2 Air sensors

As their name suggests, these collectors produce hot air and are therefore ideal for certain ventilation, soft ventilation and space heating installations. In the case of air heating, the air to be heated can be passed directly through the collector.

One particular application for these sensors is hay drying. They are lightweight and have no cooling or boiling problems, which is an advantage over liquid sensors.

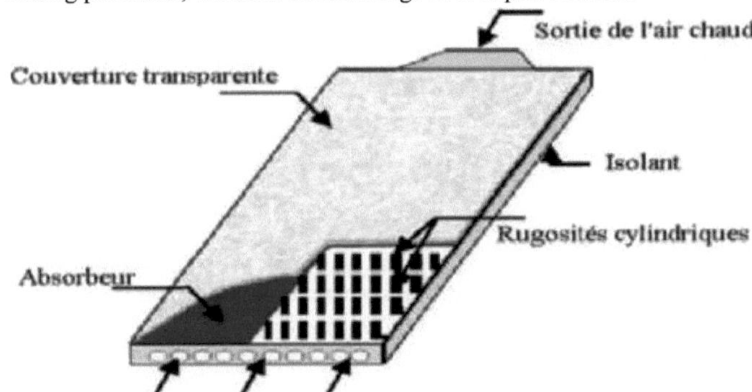

Fresh air inlet

Figure II.8: Flat air collector

II.3 Construction elements of a planar sensor

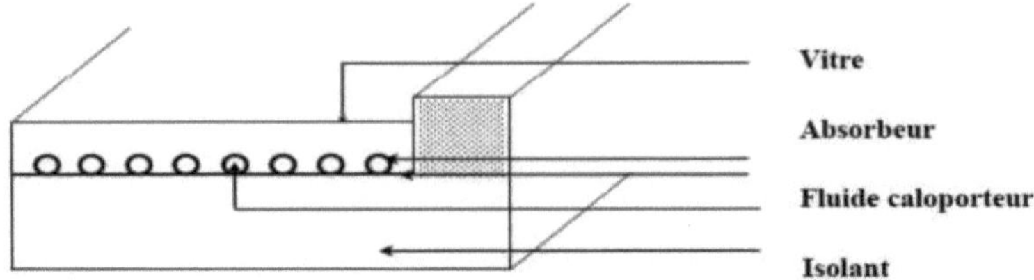

Figure Π.9: Components of a glazed flat-plate solar collector

11.3.1 The box

The casing forms the rear and side envelope of the device, and is made of selected materials, such as sheet or profiled metals, reinforced plastic materials and even counterplate.

The most commonly used metals are galvanised or pre-lacquered galvanised steel and aluminium alloys. [9]

There are two box structures:

11.3.1.1 The simple box

It consists of a single layer of material in the shape of a trough, in which the insulation and absorber are mounted.

11.3.1.2 The double box

It has a box structure for greater rigidity and better integration of insulation.

11.3.2 The transparent cover

This is the part through which the radiation reaches the surface of the absorber.

One of its characteristics is that it produces the greenhouse effect.

The most commonly used materials are :

The glass.

> Polycarbonates (Lexan, Makrolon).

> Polymethyl methacrylates (Plexiglas, Altuglas).

Glass with the lowest iron oxide content (e.g. horticultural glass) is preferable.

There are also multiple transparent covers (double covers) and covers made of a flexible transparent film such as Mylar or Tedlar:

> Absorber protection and thermal insulation.

> Part of thermal insulation.

> It reflects as little radiation as possible and absorbs as little light as possible.

so that all the radiation reaches the surface of the absorber.

11.3.3 The absorber

This is the central part of the collector where the water from the storage circuit collects the heat from the sun. The functions of this part of the collector are both thermal and hydraulic:

▪S **Water film absorber**

Where the heat transfer liquid comes into contact with most of the absorbent surface.

J **Tube absorber**

Where the liquid circulates in pipes to which the solar heat is drained by a system of heat-conducting fins.

The use of tubes provides greater resistance to pressure.

The absorber should be made of a material with good thermal conductivity, usually 0.2mm thick copper or aluminium foil with variations of 0.15 →0.3mm.

A wide variety of materials are used in these two types of absorber, including galvanised steel, black and stainless steel, copper and even plastics such as black-tinted polyethylene. [9] The absorbers often used and their thermal conductivities are given in the following table:

Materials	Temperature °C	λ W m/°C m^2
Aluminium	100	205
	200	230
	400	318
	600	423
"Coppersun	100	381
"	200	372
Copper	20	393
Magnesium	20	154

Table II.1: Absorber materials

To increase its absorption coefficient, the absorber is often coated with a thin layer of selective paint. The following table compares some coating paints

Coatings	A
oil painting :	
- black	0,90
- creamy white	0,3 - 0,35
- light grey	0,50 - 0,75
- Red	0,74
aluminium paint :	
- cellulose lacquers	0,5 - 0,55
- black	0,94
- brown	0,79
- dark green	0,88
dark blue	0,91

Table II.2: Coatings

Coppersun" is most often used as an absorber. It is a corrugated copper sheet on which copper oxide has been deposited, with two differently treated sides, the light grey absorbing side being the one on which the deposit has been made, and it has a solar radiation absorption coefficient of around 96.5%.

It also has microscopic cavities that absorb solar radiation but are small enough for the surface to be considered flat.

The table below shows the properties of some selective surfaces:

Surfaces	A	ε
nickel black on nickel	0,95	0,07
chrome black on nickel	0,95	0,09
copper black on copper	0,88	0,15
iron oxide on steel	0,85	0,08

Table II.3: selective surfaces

11.3.4 The heat transfer fluid

The working fluid is responsible for transporting heat between two or more temperature sources. It is chosen on the basis of its physical and chemical properties, and must have high thermal conductivity, low viscosity and high heat capacity. In the case of flat-plate collectors, water is used, to which an antifreeze (usually ethylene glycol) is added, or air. Air has the following advantages over water [9]:

- No problem with freezing in winter or boiling in summer.
- No corrosion problems (dry air).
- Any leak is of no consequence.
- It is not necessary to use a heat exchanger for space heating.
- The system to be implemented is simpler and more reliable.

However, it has a number of disadvantages:

- The air can only be used for space heating or solar drying.
- The product of density and heat capacity is low (p.Cp=1225 J/m3. K) for air, compared with 4.2.106 J/m3. K for water.
- The pipes must have a large cross-section to allow sufficient flow.

11.3.5 Thermal insulation

The absorber must transmit the energy collected to the heat transfer fluid, avoiding heat loss

by conduction, convection and radiation from the various peripheral parts to the outside.
The following solutions are available:

- **Front part of the absorber**

The layer of air between the pane and the absorber acts as an insulator against the transmission of heat by conduction. However, if the air space is too thick, natural convection occurs, resulting in energy loss. For the usual operating temperatures of a flat-plate collector, the air gap should be 2.5 cm thick [9].

Placing two panes of glass limits losses due to re-emission as well as losses by conduction and convection [8].

- **Rear and side sections**

To limit heat loss around the periphery of the collector, one or more layers of insulation can be used, which must be resistant to high temperatures and not outgas, otherwise you can expect to see a deposit on the inside of the cover. In addition to using insulation to minimise heat loss, the contact resistance between the plate, the insulation and the casing can be increased by avoiding pressing these surfaces against each other, because in the case of high roughness, there may be a film of air between the two surfaces in contact which prevents heat from passing easily by conduction [9], [10], [11].

The following table lists some of the properties of these insulators:

designation	λ (W/m.K)	p g/m(3)	T_{max} use	Comments
glass wool	0.034 at 0°C 0.053 at 200°C	70	150	sensitive to humidity
glass foam	0,057	123	150	********
Wood chipboard sawdust	0,13 à 0,40 0,1 0,11	********	*********	*********
vermiculite	0,12 à 0,40	********	********	********
expanded cork	0,045	100	110	********
polystyrene	0,042 0,040	15 17	85 85	compressed moulded
polyurethane polyurethane	0,035 0,027	35 35 - 40	85 110	foam tablet

Table II.4: Insulation properties

II.4 Orientation and inclination of a planar sensor

11.4.1 Orientation

Because the sun's rays are distributed evenly throughout the day, the collectors need to be oriented in such a way that the maximum amount of energy is collected. Generally speaking, collectors are oriented due south (for the northern hemisphere).

11.4.2 The tilt

The problem of inclination is the most delicate and requires a detailed study, but we can see that the vertical position of the collector favours the winter period, whereas the horizontal position leads to better yields during the summer. The ideal solution would be to incline the collectors differently depending on the time of year. As the collectors will necessarily be fixed, they will be inclined to the horizontal by an angle equal to the latitude of the location. [12].

11.4.3 Sensor connection [13] ..." ...

❖ **♦♦ Connecting sensors in series**

In this type of arrangement, the output of the first sensor is connected to the input of the second sensor, whose output is connected to the input of the third sensor, and so on.
The longer the heat transfer fluid travels, the higher the temperatures obtained at the outlet of the last collector.

❖ **Parallel connection of sensors**

In this case, the water arrives at each collector via a distribution pipe that runs along the lower edges, while the hot water is drawn off via another pipe placed along the upper edge of the collector. It is therefore important in this type of connection that the circuit is well balanced so that the flow rate of the heat transfer fluid is evenly distributed throughout the various collectors. See figure (I.10).

❖ **Mixed connection**

This is an assembly that combines the series and parallel modes, resulting in a more uniform distribution of flow and therefore temperatures. There are two types of mixed connection, particularly suitable for large installations.
- Parallel series connection.
- Parallel series connection.

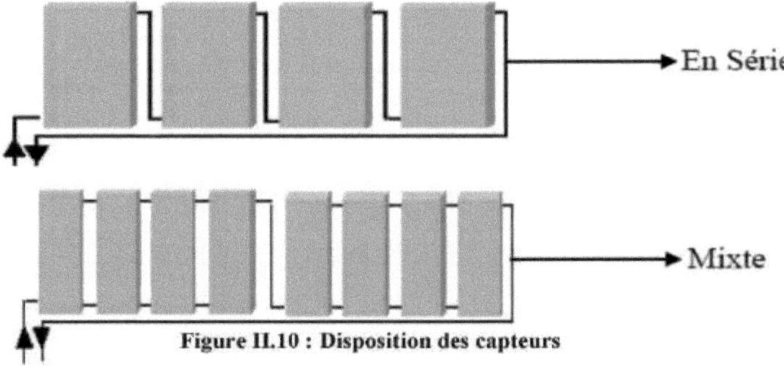

Figure II.10 : Disposition des capteurs

II.5 The storage tank

The storage tank is an essential part of a solar water heating system. As its name suggests, it is used to store the hot water coming from the collectors so that it can be released when needed. It may or may not contain a heat exchanger. See Figure (II.11).
To prevent heat loss to the outside environment, the storage tank must be well insulated, with the right thickness of thermal and economic insulation.
Storage is characterised by [14]:

❖ Heat accumulation mode (sensible or latent).
❖ The thermal capacity of the stock.
❖ Heat loss from storage.

Depending on the storage capacity, there are two types:

❖ Long-term storage (inter-seasonal).

*** Short-term storage (no more than a few days or hours).

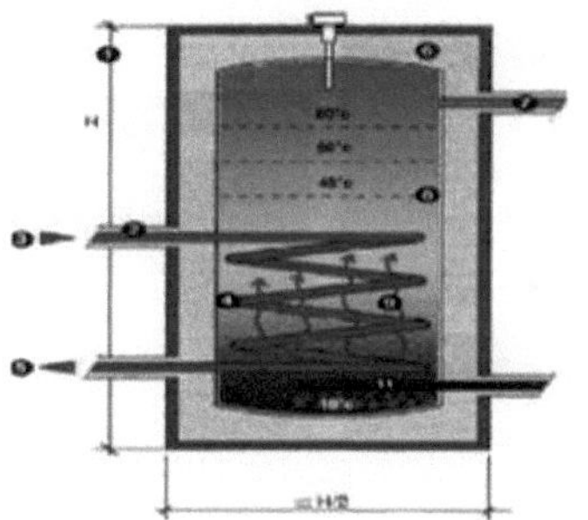

1 : Hauteur
2 : Isolation
3 : Arrivée chaudière
4 : Circuit primaire
5 : Retour chaudière
6 : Isolation (8 à 16 cm)
7 : Sortie eau chaude
8 : Surface d'éch. min entre eau fr. et ch.
9 : Ech. circuit primaire et eau sanitaire
10 : Entrée eau froide
11. Casse jet, réduit vitesse arrivée eau

Figure II.11: The storage tank with heat exchanger

II.6 Piping

The pipework is used to transfer the heat transfer fluid, and its design and installation must be followed carefully to avoid serious problems.

The pipe circuit must be as simple as possible, i.e. short and avoid changes in diameter.

The pipes must be carefully insulated. The heat transfer fluid circuit comprises several devices, the main components of which are.

- A safety valve.
- An expansion vessel.
- A steam trap.
- A non-return valve.
- Insulation.

II.7 The heat exchanger

A heat exchanger is a device in which two fluids, separated by a wall, circulate and exchange heat. One fluid cools down while the other heats up.

It is usually built into the storage tank, but can also be located outside.

Because of its heat transfer function, the heat exchanger must offer the largest possible contact surface between fluids, which is why most heat exchangers have the appearance of a coil.

Note: The external exchangers will be plate type exchangers and the internal exchangers will be coil type, in our installation the exchanger used is coil type. Figure (II.12). shows the two types of exchanger.

Echangeur à plaque

Echangeur intégré

Figure II.12 : Types of heat exchanger

II.8 How a solar water heater works

It works simply by transferring the solar energy absorbed by the collectors (heat) to a storage system (tank).
The transfer is carried out by means of a heat transfer liquid.
The heat-transfer liquid must move from the solar collector to the storage tank (where it exchanges its heat to heat the cold water contained in the tank), the cooled water returning to the collector where it will be reheated by solar radiation. See Figure (**II**.13).
During its journey from collector to reservoir to collector, the water can either circulate by itself (natural circulation) or be driven by a small pump (forced circulation).
The relative positions of the isolator and the reservoir are essential in deciding how the assembly will function [15], [16].

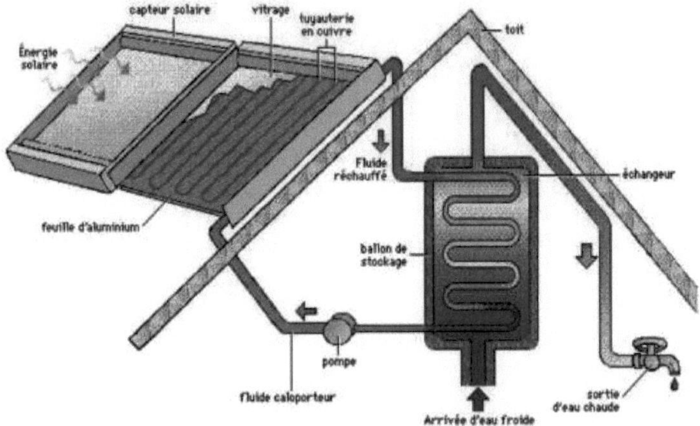

Figure II.13: Block diagram of a solar water heater.

II.9 Types of solar water heaters

II.9.1 Natural circulation solar water heater (thermosiphon)

The water heated in the collector is sent to the storage tank, where it is replaced in the collector by cold water, which in turn heats up, and so on. 9] The circulation of water in the circuit must be continuous as long as there is sunshine and water to heat, and is generally fairly slow.
The hot water, which is lighter than the cold water, first accumulates at the top of the tank. See Figure (**II**.14).

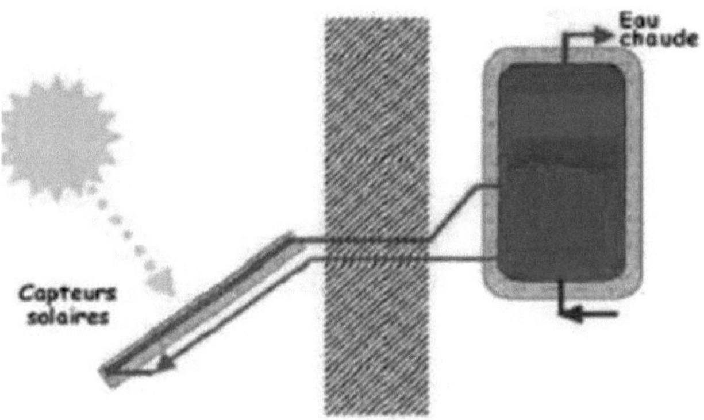

Figure II.14 : The natural circulation solar water heater without heat exchanger

II.9.2 Natural circulation solar water heater with heat exchanger

The hot water coming from the collector circulates inside the tank in a heat exchanger. When it comes into contact with the cold water in the tank, it gives up its calories through the wall of the exchanger and returns to heat up in the collector. See Figure (I.15).

The water heated in the collector remains in a closed circuit called the "primary circuit". Heated water in contact with the heat exchanger rises to the top of the tank, while cold water descends and is reheated at the bottom of the tank. [16]

On the subject of freezing and scaling, we will see the advantages of using an Exchanger.

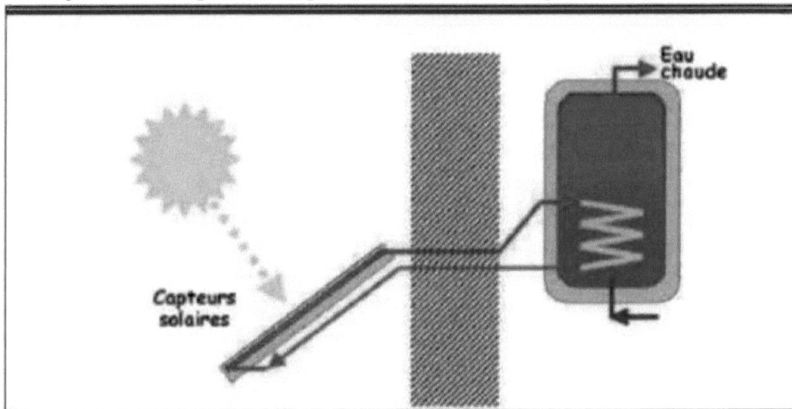

Figure II.15 : Thermosiphon solar water heater with heat exchanger

11.9.3 Solar water heater with circulator and exchanger inside the house

The fluid circulating in the primary circuit (collector - exchanger - collector) is generally different from the water stored in the tank, so a certain heat-transfer fluid will circulate in the primary circuit, absorbing heat energy inside the collector and then releasing it to the atmosphere.

See Figure (**II**.16). [9]

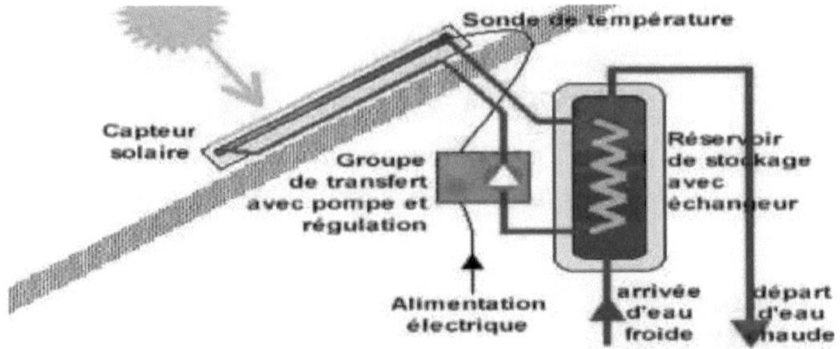

figure II.16: The water heater with circulator and heat exchanger inside the storage tank

II.9.4 Solar water heater with circulator and exchanger outside the storage tank.

This installation consists of three main loops through which water or an antifreeze fluid circulates, as shown in Figure (Π.17). [16]

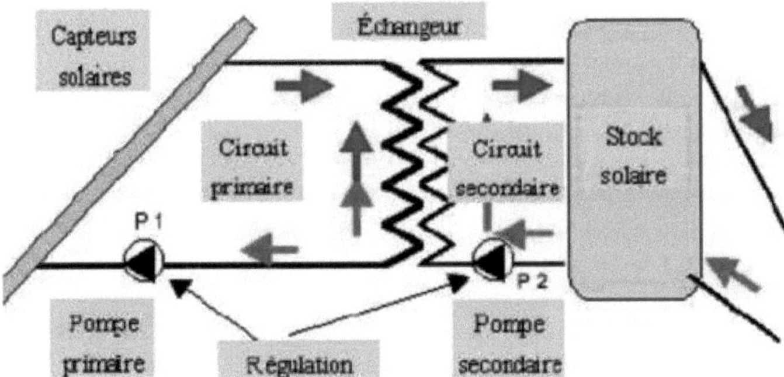

Figure II.17: Water heater with circulator and heat exchanger outside the tank storage

II.9.4 .1 The primary loop

This is a closed loop between the collectors and the heat exchanger. The fluid is heated in the solar collectors and then cools by transferring its heat to the secondary loop in the heat exchanger.

Circulation takes place when the solar energy available is sufficient to heat the water in the collectors. The regulation system controls the start and stop of the primary pump according to the amount of sunshine detected by temperature sensors in the primary and secondary circuits.

II.9.4.2 The secondary loop

This is an open loop between the heat exchanger and the solar storage tank, with domestic hot water circulating in this loop.

The solar storage water is heated in the heat exchanger and cools down when hot water is used. The water temperature is slightly lower than in the primary loop, with a maximum of around 90°C.

Circulation takes place when the primary circuit is able to heat the secondary circuit. The control system controls the start and stop of the secondary pump according to the

temperatures in the primary and secondary circuits and the state of the primary pump.
The distribution loop is an open loop, with cold water from the mains entering at the bottom of the tank and hot water leaving at the top.

Conclusion

Solar installations can be used in all climates to produce hot water, but their annual performance is proportional to the amount of sunshine in the area where the solar collectors are installed.
The choice between types of solar collector is determined by the type of application required, reliability, price and desired temperatures.

CHAPTER 3

THERMAL BALANCE OF A FLAT-PLATE SOLAR COLLECTOR

III.1 Background on heat transfer

Thermodynamics makes it possible to predict the total amount of energy that a system must exchange with the outside world in order to move from one state of equilibrium to another. Thermodynamics (or thermokinetics) aims to describe quantitatively (in space and time) the evolution of the characteristic variables of the system, in particular the temperature, between the initial state of equilibrium and the final state.

Heat flows under the influence of a temperature gradient by conduction from high to low temperatures. The amount of heat transmitted per unit time and per unit area of the isothermal surface is called the heat flux density.

III.1.1 Conduction

It is the transfer of heat within an opaque medium, without the displacement of matter, under the influence of a temperature difference. The propagation of heat by conduction within a body takes place by two distinct mechanisms: transmission by the vibrations of atoms or molecules and transmission by free electrons.

The theory of conduction is based on Fourier's hypothesis: the flux density is proportional to the temperature gradient:

$$Q_{cd} = -\lambda . S . \overrightarrow{grad}(T) \qquad \text{(III.1)}$$

$$Q_{cd} = -\lambda . S . \frac{\partial T}{\partial x} \qquad \text{(III.2)}$$

With :

Qcd: heat flow by conduction (W).

S: Heat flow cross-sectional area (m2).

λ: Thermal conductivity (W/m°C).

x: Space variable in the direction of flow (m).

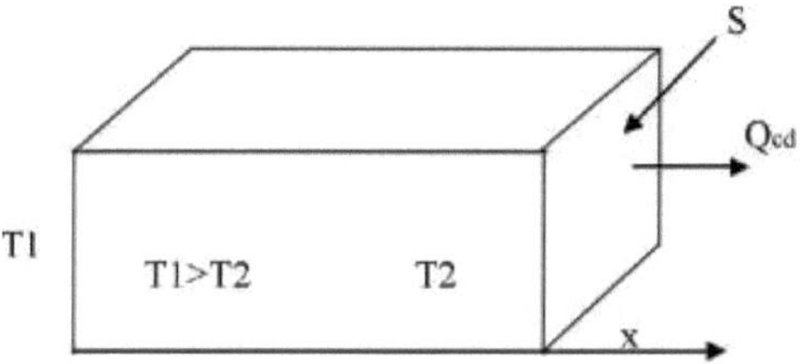

Figure III.1: Diagram of heat transfer by conduction

III.1.2 Convection

In this case, heat is transferred from a liquid or gaseous fluid to a solid body (for example between air and a wall). The particles are in motion between themselves.

There are two types of convection:

III.1.2.1 Free or natural convection

Fluid movement is generated by variations in density caused by variations in temperature within the fluid, as in the case of thermal circulation.

III.1.2.2 Forced convection

The movement of the fluid is induced by a cause independent of temperature differences (pump, ventilation, etc.).

This is the transfer of heat between a solid and a fluid, the energy being transmitted by the movement of the fluid.

This transfer mechanism is governed by Newton's law:

$$Q_{cv} = h_c .S.(T_p - T_\infty) \qquad \text{(III.3)}$$

Q_{cv} : convection heat flux (W).

S: Heat transmission surface area (m2).

Tp: Solid surface temperature (°C).

T_z : The temperature of the fluid before it comes into contact with the solid (°C).

h_c : Convective heat transfer coefficient (W/m2.°C).

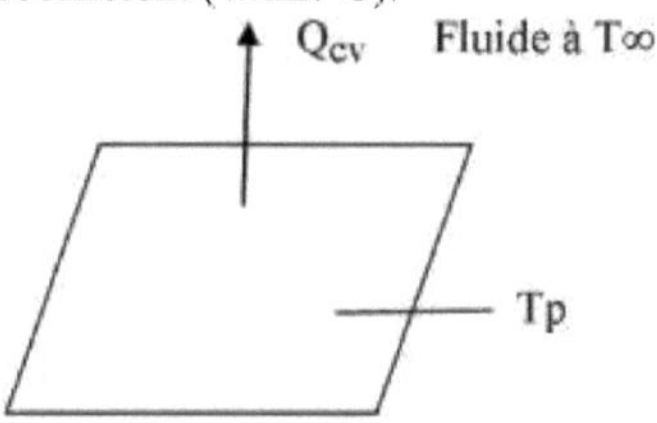

Figure III.2: Diagram of heat transfer by convection

III.1.2.3 Calculates the convection exchange coefficient

Forced convection :

In the absence of natural convection, the exchange coefficient hc by convection is independent of the difference in temperature between the wall and the fluid, but depends on the following 6 variables:

U_m average fluid velocity

p density of the fluid

Cp specific heat of the fluid

μ thermal viscosity of the fluid
λ thermal conductivity of the fluid
D characteristic dimension of the exchange surface
From these quantities, we define the following dimensionless numbers:

$$Nu = hc\frac{D}{\lambda}$$ Nusselt number λ

$$\mathrm{Re} = \frac{\rho.U_m.D}{\mu}$$ Reynolds number

$$\mathrm{Pr} = \frac{\mu.Cp}{\rho}$$ number of Prandts

Experimental work studying heat transfer by convection in a situation where
Data provide their results in the form of mathematical correlations Nu=f(Re,Pr) which can be used to calculate hc by :

$$h = Nu\frac{D}{\lambda} \qquad \text{(III.4)}$$

Re: the Reynolds number characterises the fluid flow regime
Pr: the Prandtl number characterises the heat exchange between the fluid and the wall.
Nu: the Nusselt number characterises the heat exchange between the fluid and the wall.

Natural convection :

In natural convection, the movement of the fluid is due to variations in the density of the fluid resulting from heat exchange between the fluid and the wall. The fluid is set in motion by Archimedes' forces because its density is a function of its temperature.
Forced convection is negligible if

$$Gr/P_r^{\,2} > 100$$

$$Nu = C\,(Gr\ Pr)^n$$

Laminar convection Gr Pr < 109 => n=1/4
Turbulent convection Gr Pr > 109 =>n=1/3

III.1.3 Radiation

Heat transfer by radiation occurs when energy in the form of electromagnetic waves is emitted by one surface and absorbed by another. This exchange can take place when the bodies are separated by a vacuum or by any intermediate medium that is sufficiently transparent for electromagnetic waves.

The fundamental law of radiation is Stefan-Boltzmann's law:

$$Q_r = \varepsilon.\sigma.s\left(T_p^4 - T_\infty^4\right) \qquad \text{(III.5)}$$

Q_r : heat flux density emitted by the body.
ε: thermal emissivity of the material.
σ: Stefan-Boltzmann constant evaluated at 5.6.10 8 W/m 2K 4

Tp: Surface temperature.
T∞: Temperature of the medium surrounding the surface.
S: Area of the surface
Surrounding environment at T∞

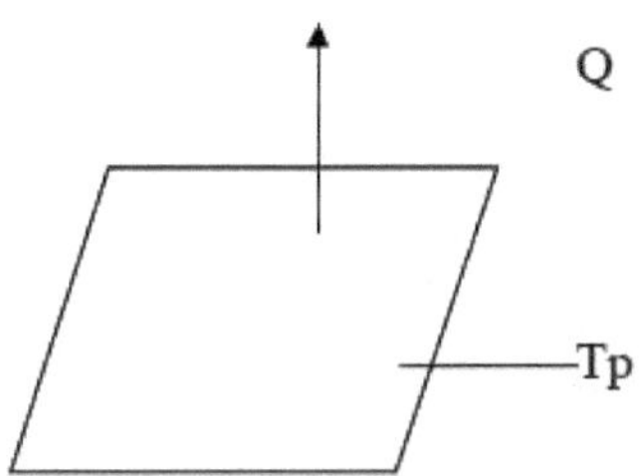

Figure III.3: Diagram of radiative heat transfer.

III.2 The different modes of heat transfer in a solar collector

A solar collector simultaneously involves the three modes of heat transfer, conduction, convection and radiation (Figure III.4).

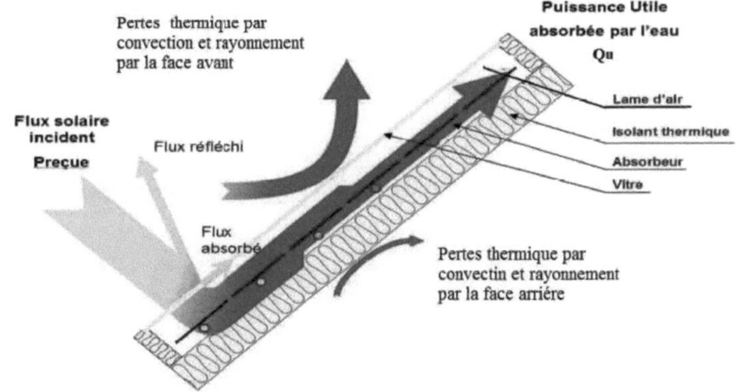

Figure III.4 The various heat exchanges in a flat-glass collector

III.2.1 Heat exchange at the glass pane [16]

In addition to the heat flux exchanged by radiation between the absorber and the glazing, there is another incident heat flux (Qv) which will be absorbed by the pane but of little importance.

$$Q_v = \alpha_v .Sv.G \qquad (III.6)$$

α_v : absorption coefficient of the pane.
Sv: surface area of the pane (m2).
G: Global radiation.

> The heat flow exchanged by convection between the glazing and the ambient air is given by equation (III.7)

$$Q_{cvam} = h_{cvam} \cdot Sv \cdot (Tv - Tam) \qquad (III.7)$$

Tam: Ambient temperature (°K).
h_{cvam} : convection exchange coefficient between the pane and the ambient air. This is due to

entirely at wind speed (W/ m2 °K).
The Hottel-Woertz correlation can be used.

$$h_{cvam} = 5.67 + 3.86 \cdot V \qquad \text{(III.8)}$$

V: Wind speed (m/s).
> The heat flux exchanged by radiation between the glazing and the sky is given by equation (III.9)

$$Q_{rvc} = h_{rvc} .Sv.(Tv - Tciel) \qquad \text{(III.9)}$$

T i_{ce} l: Ceil temperature (°K).
h_{rvc} : radiation exchange coefficient between the pane and the sky (W/ m2°K).

$$h_{rvc} = \varepsilon v.\sigma.(Tciel - Tv).(Tv^2 + Tciel^2) \qquad \text{(III.10)}$$

$$Tciel = 0.0552.Tam^{1.5} \qquad \text{(III.11)}$$

III.2.2 Heat exchange at the absorber [16]
> The heat flux exchanged by convection between the absorber and the glazing is given by Equation (III.12)

$$Q_{cabv} = Sab.h_{cabv} .(Tab - Tv) \qquad \text{(III.12)}$$

Tab: Absorber temperature (°K).
Sab : Absorber surface area (m2)
h_{ca} bv: The convective heat exchange coefficient between the pane and the absorber (W/m2°K).

$$h_{cabv} = N_u \frac{Kair}{b}$$

$$Nu = 1 + 1.44\left(1 - \frac{1708}{Gr.Pr.cos\beta}\right)\left(\frac{|x| + x}{2}\right)\left(\frac{|y| + y}{2}\right)$$

$$x = 1 - \frac{1708sin(1.8\beta)}{Gr.Pr.cos\beta}$$

$$y = \frac{(Gr.Pr.cos\beta)^{\frac{1}{3}}}{5830}$$

***1 (Gr. Pr. cosβ)* a 5830**
r: Prandtl number.
b: thickness of the air gap separating the pane from the absorber plate
(m). Gr: Grashof number.
Kair: thermal conductivity of air (W/m.K).
β: inclination of the sensor (rad).
> The heat flux exchanged by radiation between the absorber and the glazing is given by equation (III.15)

$$Q_{rabv} = S_{ab}.h_{rabv}.(T_{ab} - T_v) \qquad \text{(III.15)}$$

$$h_{rabv} = \frac{\sigma.(T_v + T_v)(T_{ab}^2 + T_v^2)}{\frac{1}{\varepsilon_{ab}} + \frac{1}{\varepsilon_v} - 1} \qquad \text{(III.16)}$$

εab, εv: are the emissivities of the absorber and glazing respectively.
σ: Stéphane Boltzman constant σ = 5.67 10 8 W/m2 K4.
The heat flux exchanged by conduction between the absorber and the insulator is given by equation (III.17)

$$Q_{cdabi} = \frac{(T_{ab} - T_i)}{\left(\frac{L_{ab}}{S_{abi}.\lambda_i}\right) + \left(\frac{L_i}{S_{abi}.\lambda_{ab}}\right)} + \frac{(T_{ab} - T_i)}{\left(\frac{L_{ab}}{S_{abil}.\lambda_{il}}\right) + \frac{L_{il}}{S_{abil}.\lambda_{ab}}} \qquad \text{(III.17)}$$

Ti: insulation temperature (°K).
Sabil: absorber-insulator contact surface for the side face (m2). λi: thermal conductivity of the insulation (rock wool). (W m 1 °K 1). λil: thermal conductivity of the insulator (glass wool). (W m 1 °K 1).
λab: thermal conductivity of the absorber (W m 1 °K
I: . Lab: absorber thickness (m).
II: thickness of insulation (rock wool) (m).
III: thickness of side insulation (glass wool) (m).
> The heat flux exchanged by convection between the absorber and the heat transfer fluid (water) is given by equation (III.18)

$$Q_{cabf} = h_{cab}.Sabf.(Tab - Tf) \qquad \text{(III.18)}$$

Tf: temperature of the heat transfer fluid (°K).
Sabf: heat transfer fluid absorber contact surface (m2).
h_{ca} bf coefficient of exchange by convection between the absorber and the heat transfer fluid (W/ m2 °K).
The convective heat exchange coefficient inside the tubes h_{ca} bf is calculated as follows following Gnielinski. He used a large number of experimental data on the transfer of heat in the tubes and proposed a correlation that could be used for the transition regime and for the turbulent regime, taking into account the flow establishment length. The physical properties are calculated at the average water temperature.

$$N_u = \frac{\Omega}{8}\frac{(R_e - 10^3)P_r}{1 + 12.7\left(\frac{\Omega}{8}\right)^{0.5}\left(P_r^{\frac{2}{3}} - 1\right)}\left[1 + \left(\frac{d_i}{l}\right)^{\frac{2}{3}}\right] \qquad \text{(III.19)}$$

Ω: Darcy coefficient.
Pr: Prandtl number.
Re: Reynolds number. di: inside diameter (m).
L: tube length (m).
This correlation can be used for 0.6 < Pr < 2000, 2300 < Re < 10^6 . The Reynolds number is

given by:

$$\mathrm{Re} = \rho \frac{D.V}{\mu} \qquad \text{(III.20)}$$

D: pipe diameter (m).

μ: dynamic viscosity of water (Pas).
V: average fluid velocity (m/s).

For a smooth hydraulic turbulent flow, the Darcy coefficient is given by different relationships, depending on the Reynolds number.

Si $2300 \leq \mathrm{Re} \leq 10^5$ apply the Blasius formula:

$$\Omega = 0.3164\mathrm{Re}^{-0.25} \qquad \text{(III.21)}$$

Si $10^5 \leq \mathrm{R} \leq 10^6$ we apply Herman's relation :

$$\Omega = 0.0054 + 0.3964\mathrm{Re}^{-0.3} \qquad \text{(III.22)}$$

The internal heat exchange coefficient is given by:

$$h_{cabf} = Nu \frac{\lambda eau}{di} \qquad \text{(III.23)}$$

λwater: thermal conductivity of water (W m^1 °K 1) d: diameter of inner pipe (m).

> The incident heat flux received by the absorber is given by (III.24) :

$$Q_{ab} = \alpha_{ab} . \tau_v .Sab.G \qquad \text{(III.24)}$$

α_a b: absorption coefficient of the absorber.
τ_v transmission coefficient of the pane.
G: global illuminance incident on the inclined plane of the flat-plate collector. (W/m2).

III.2.3 Heat balance of the flat plate solar collector in transient operation [16].
The total balance, which gives the thermal behaviour of the collector, and provides the ntiel average orditemperatures of the absorber, the glass and the heat transfer fluid, is given by system of non-linear differential equations:

Glass

$$m_v c_v \frac{dT}{dt} = \alpha_v . s_v . G + S_{ab}(h_{cabv} + h_{rabv})(T_{ab} - T_v) - h_{cvam} . s_v . (T_v - T_{am}) - h_{rvc} . s_v . (T_v - T_{ciel}) \qquad \text{(III.25)}$$

Absorber

$$m_{ab} . c_{ab} . \frac{dT_{ab}}{dt} = \alpha_{ab} . \tau_v . s_{ab} . G - s_{ab}(h_{cabv} + h_{rabv})(T_{ab} - T_v) - (\varphi_1 + \varphi_2)(T_{ab} - T_i) - h_{cabf} . s_{abf}(T_{ab} - T_f) \qquad \text{(III.26)}$$

- **Heat transfer fluid**

$$m_f.c_f.\frac{dT_f}{dt} = h_{cabf}.s_{abf}(T_{ab} - T_f) = Q_u \quad (III.27)$$

$$\varphi_1 = \frac{1}{\left(\frac{L_{ab}}{s_{abi}.\lambda_i}\right) + \left(\frac{L_i}{s_{abi}.\lambda_{ab}}\right)} \qquad \varphi_2 = \frac{1}{\left(\frac{L_{ab}}{s_{abil}.\lambda_{il}}\right) + \left(\frac{L_{il}}{s_{abil}.\lambda_{ab}}\right)} \quad (III.28)$$

With :

mv, mab, mf: respective masses of glass, absorber and fluid.

Cv, Cab, Cf: heats of mass of the glass pane, absorber and heat transfer fluid respectively.

III.3 Overall energy loss [16]

Heat losses are due to the difference in temperature between the absorber and the surrounding environment. They occur in the three modes of heat transfer. They are divided into three categories: forward losses, backward losses and lateral losses.

III.3.1 Coefficient of heat loss to the front of the collector

The overall heat loss coefficient towards the front of the collector is given by the following relationship:

$$U_{av} = \frac{1}{\left(\frac{1}{h_{rvc} + h_{cvam}}\right) + \left(\frac{1}{h_{cabv} + h_{rabv}}\right)} \quad (III.29)$$

III.3.2 Coefficient of heat loss to the rear of the collector

This coefficient is not as high as before, as the sensor is very well insulated at the rear.

The expression evaluating this coefficient is given by:

$$U_{arr} = \frac{K_{isol}}{E_{isol}} \quad (III.30)$$

Kisol: thermal conductivity coefficient of the insulation (W/ °K m).

Eisol: insulation thickness (m).

III.3.3 Lateral heat loss coefficient

The value of this coefficient is lower than that of the rear loss coefficient, given that the lateral surface of the collector is not very large.

$$U_{lat} = \left(\frac{K_{isol}}{E_{isol}}\right)\left(\frac{A_{lat}}{A_c}\right) \quad (III.31)$$

Alat: lateral surface area of the collector (m2).

Ac: collector surface area (m2).

The overall external heat loss coefficient is the sum of the three coefficients.

$$UL = Uav + Uarr + Ulat \quad (III.32)$$

III.4 Instantaneous solar collector efficiency [16]

The analysis carried out in this field by Hottel, Willier, Wortz and Bliss leads to a single equation giving the instantaneous efficiency of the collector, which is defined by the following ratio:

Output power = Captured power - Losses (III.33)

η = Useful thermal power per m2of collector / solar flux incident on the collector surface

$$\eta = \frac{Qu}{Ac.G} = \frac{Ac[(\alpha\tau)eff.G - UL(Tab - Tam)]}{Ac.G} \quad (III.34)$$

Qu: useful power recovered by the heat transfer fluid (W).

$\alpha\tau$: are respectively the absorption coefficient of the absorber and the transparency of the glazing. * The instantaneous efficiency of the collector as a function of mass flow rate is given by [17]:

$$\eta i = m . Cp . (Tfs - Tfe)/Ac . G \quad (III.35)$$

With,

m: heat transfer fluid mass flow rate

Cp: specific heat of water.

Tfs: fluid outlet temperature.

Tfe: fluid inlet temperature.

Ac: surface area of solar collector.

G: global incident solar flux.

III .5.simulation principle:

In our 1[er] case we assume that the collector surface is A=2.75m^2 ,number of collectors N=1, mass flow rate m=0.05kg/s,Cp(water))4.190kj/kg,Cp(air)=1kj/kJ, to determine the output temperature (determining the influence of the type of collector on the output temperature), and then using the same initial condition, we want to draw up a comparison in terms of optimisation between a flat plate collector and an evacuated tube collector, often under the same conditions, and extrapolate which collector has the best instantaneous yield for a solar water heating installation. To obtain the required results we need to use an energy simulation software package called tyrnsys, which is a flexible environment for carrying out simulations and optimisations using clearly defined diagrams.

At present, solar cooling simulation codes generally use steady-state models to describe the behaviour of evacuated collectors. This efficiency is generally presented in the quadratic form proposed by European standards (CEN, 2001) [18] :

where :

$$\eta = \eta_0 - U_1 \frac{T_{mf} \ T_e}{G^*} - U_2 . \frac{(T_{mf} \ T_e)^2}{G^*}$$

= is the optical efficiency of the sensor, with approximately 0.792 for a and 0.762 for a vacuum tube collector. [19]

- *Tmf* = is the average temperature of the heat transfer fluid.
- *Te* = is the outdoor temperature near the sensor
- *G** = is the global solar radiation in Wh/m
- U_1 and *U2* = are the coefficients of heat loss by conduction and by

convection, in W/(m^2 K) with default values shown in the table below:

Type of sensor	Ui	U2
Planar sensor	6.65	0.06

Vacuum tube sensor	0.2125	0.0167

Table III.1 Heat loss coefficients.

IV I.6 Technical and economic study of a solar water heater:

We will calculate the total price of the installation by defining the price of each element of this installation (for 2 people) as follows:

@ **2.57m** vacuum tube flat plate collector² and **20 years** warranty, **price=76137.00DA**.

@ Circulator of a solar installation (pump) **price=14262.00DA.**

@ **200l** domestic hot water tank, **price=8500.00DA.**

Regulation system (the one that controls the triggering of the pump) **price=5412.00DA.**

@ **8m** from the pipe, **price=2800.00DA.**

@ ***THE FINAL PRICE OF THE INSTALLATION =107111.00DA.***

CHAPTER 4

Simulation and interpretation The results

V V.1 Introduction:

Dynamic thermal simulation makes it possible to "virtually bring the situation to life" over a long period, in order to study the predicted behaviour and obtain results that are close to reality. To validate the solutions selected, the TRNSYS software was chosen for the simulation because of its various advantages.

VI .2 TYRNSYS interface:

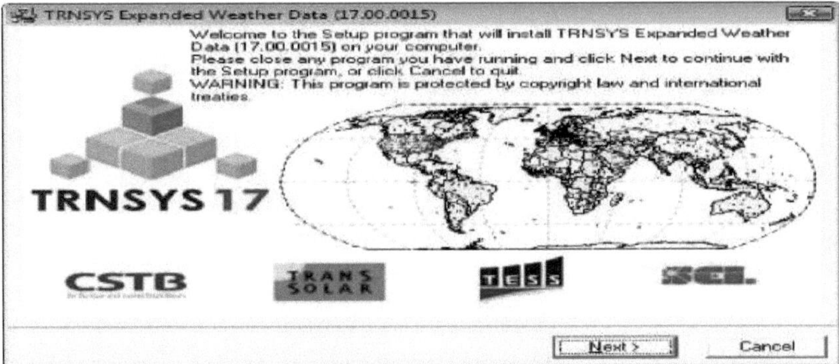

Figure (IV.1): software installation interface (TYRNSYS).

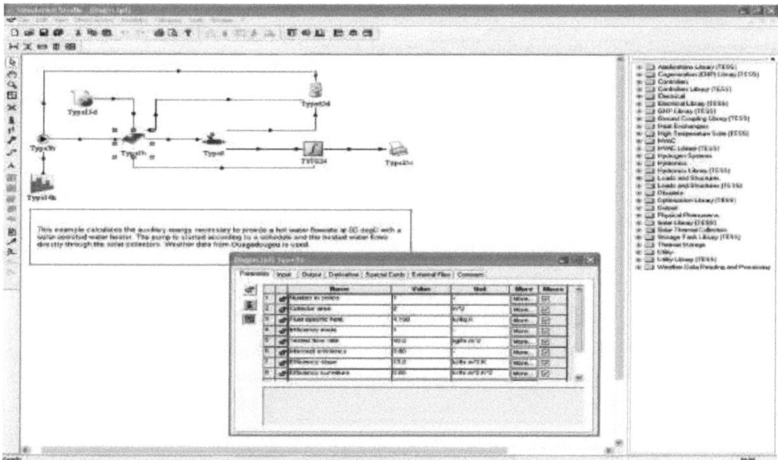

Figure (IV.2): Work plan interface (TYRNSYS).

IV.2 TRNSYS software presentation:

TRNSYS is a dynamic simulation environment that enables the behaviour of complex thermal systems to be finely simulated. **TRNSYS** has been available since 1975. It has inspired numerous developments of other simulation software, which use either its generic solver or some of its models, or both (Energy 10, Energy+, CA-SIS, HVACSIM+, etc.).

TRNSYS is based on a block diagram approach. This modular approach means that complex problems can be broken down into several less complex problems, and that work can be carried out in an 'open' environment, allowing new components and concepts to be added. A **TRNSYS** simulation project therefore consists of choosing a set of mathematical models of physical components (either based on existing models in the **TRNSYS** model libraries, or by creating them) and describing the interactions between these models. The IISiBat 3 graphical environment assists the user in these two steps with a model editor and a project editor. Each icon in an IISiBat project window represents a sub-program (traditionally written in **FORTRAN**, although **TRNSYS** version 17 allows the use of any programming language capable of generating a Windows DLL - C, C++, etc.). Each of these black boxes has a set of input variables and a set of output variables. Connecting the icons means creating connections between these variables. The user can add "boxes" and define new algorithms to simulate the behaviour of new types of objects that do not exist in the standard version of TRNSYS. This is possible by specifying equations directly (without using a programming language), by assembling existing models into 'macro models', by deriving new models from existing models by extension or by creating entirely new models, using a programming language such as FORTRAN, C or C++, or even external applications such as the generic equation solver EES.

TRNSYS 17 also contains a large number of standard models (Utilities, Storage and Thermal, Equipment, Loads and Structures, Heat Exchangers, Hydraulics, Controllers, Electrical Components/Photovoltaics, Solar Collectors, etc.). Simply Interconnect them in a project editor to define a simulation project. TRNSYS also benefits from an active community of users, who develop "freeware" models, accessible free of charge to all users.

Several TRNSYS model libraries also exist as additional commercial products, developed by specialist design offices. Several building models are available. The most comprehensive model can be used to simulate solar water heater installations and the thermal behaviour of a multi-zone building in great detail (ambient temperature, energy requirements, air humidity for each zone and surface; gains through infiltration/ventilation, convective coupling with other zones; variation in sensible energy; latent energy requirements; solar energy entering through windows; comfort; etc.).

IV.2.1. Advantages of TRNSYS software :

TRNSYS software offers many advantages:

In fact, using the simulation studio utility makes it easier to simulate a solar installation and allows the various parameters to be changed very easily. In addition, TRNSYS is a modular software package to which modules written in Fortran, Matlab or EES can be added. This means that the model can be improved by including different thermo-aerodynamic phenomena and by including a model simulating heating, air-conditioning, ventilation and refrigeration systems. Finally, and in comparison with CFD, TRNSYS is faster in simulations. Admittedly, it's not the same concept or the same equations to solve, but in our case we need to find the thermal loads of an arena, which TRNSYS can at first sight determine.

IV.3 TRNSYS environment tools :

Any study or application of solar energy at a given site requires complete and as detailed as possible knowledge of the site's insolation. This is achieved by the presence of a meteorological measuring station that has been operating regularly for several years. If there is no such station, certain approximate methods will be used to predict the characteristics of the solar radiation.

IV.3.1 Météonorm :

The METENORM software provides TRNSYS with reliable climate data for every hour of the day and for a whole year. If you don't have a weather station, METENORM can calculate the climatic conditions of a location by interpolation between different stations.

IV.3.1 Data:

A data file was selected from the database of our digital code. The data comes from daily meteorological measurements taken to obtain an annual file with an hourly time step.

This file relates to a given location: it contains the following information:

- ❖ The number of the day in each month in which the calculation is made.
- ❖ The calculation step.
- ❖ The latitude of the middle.
- ❖ The orientation of the collector in relation to the south.
- ❖ Solar flux.
- ❖ The albedo coefficient
- ❖ The glazing's transmission and absorption coefficients.
- ❖ The heat capacity of the heating fluid.
- ❖ Pump power.
- ❖ The flow rate of the heating fluid through the collector.
- ❖ The water flow rate.
- ❖ The temperature surrounding the storage, the ambient temperature and the mains temperature.

IV.3 Diagram of a solar water heater installation:

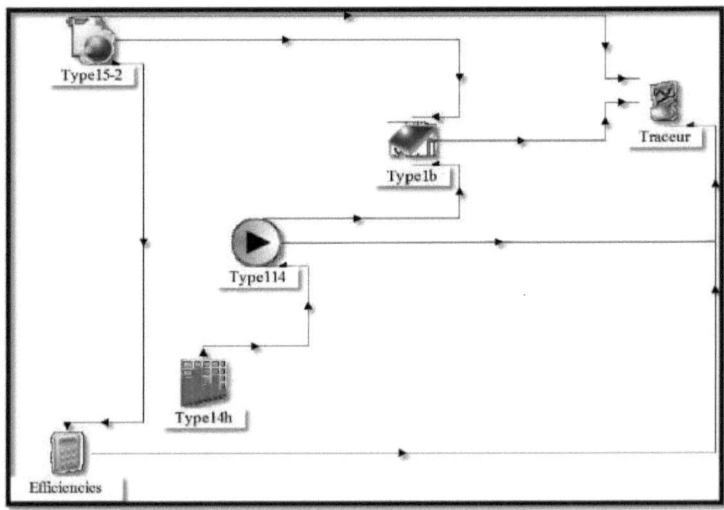

Figure (IV.3): diagram of the installation of a flat plate solar water heater.

vitré(TYRNSYS)(adrar station).

IV.4 The various components of the installation:

> Weather data.

> Pump.

> Flat plate solar collector and evacuated tube collector.

> Plotter (display of results).

A water source controls the operation of the pump (period from

IV.5 Comparison between a flat plate solar collector and an evacuated tube solar collector.

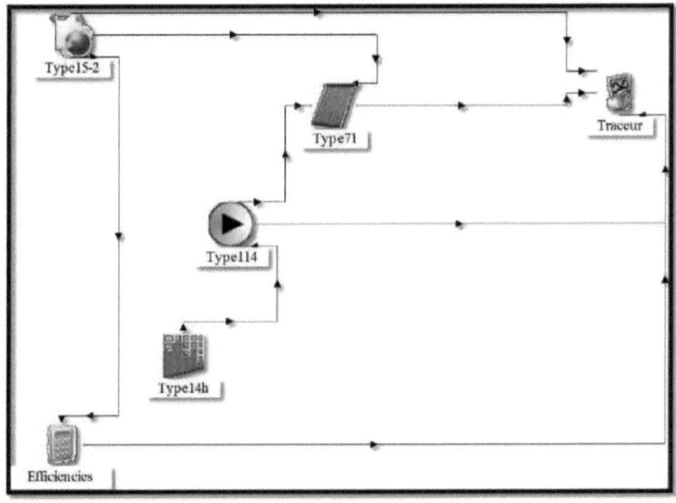

Figure (IV.4): diagram of the installation of a vacuum solar water heater (TYRNSYS) (adrar station).

- **Case 1**: The only variable that is easily comparable between simulations and measurements is the temperature of the fluid at the collector outlet. In this section, we will study the variation in the outlet temperature of the heat transfer fluid for the two types of sensor. In short, we want to see clearly which sensor gives us the best results in terms of outlet temperature.
- **a simulation within the first 10 days of February**

Figure (IV.4): shows the evolution of the outlet temperature over the period (744h,984h). It can be seen that the temperature starts to increase step by step during the hot water production period up to 60°C.

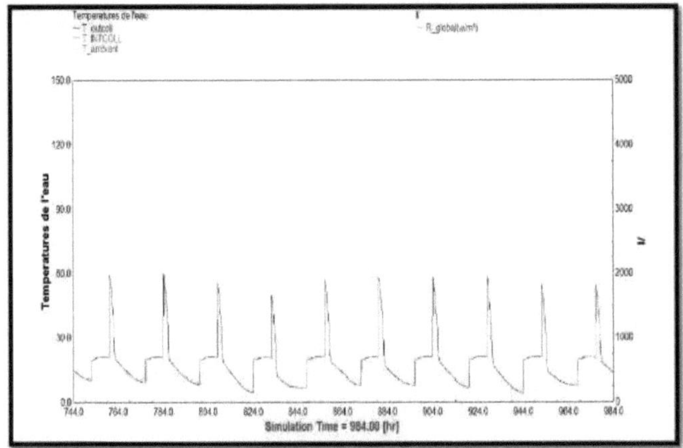

Figure (IV.5) shows the evolution of the outlet temperature over time in the case of a glazed flat-plate solar collector.

glazed flat plate solar collector.

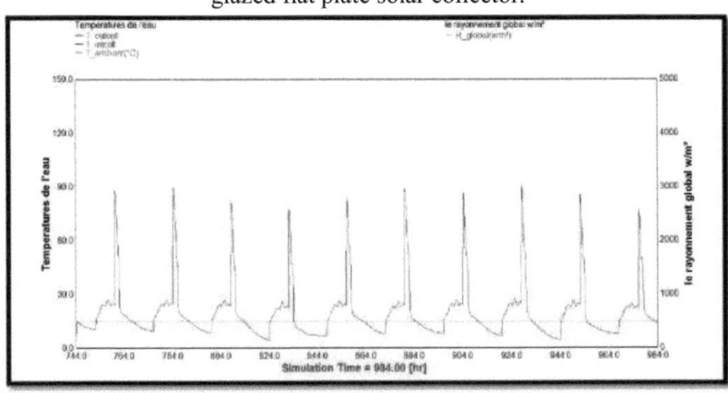

Figure (IV.6) shows the evolution of the outlet temperature over time in the case of an evacuated tube solar collector.

evacuated tube solar collector.

>Figure (IV,5) shows that the maximum temperature does not exceed 60°C during the operation of the glazed flat plate collector under the same conditions (station, flow rate, surface area, period). On the other hand, Figure (IV,6) shows that the evacuated tube flat plate collector has high output temperature values; it reaches 150°C, which confirms the performance of this type of collector for solar water heating.

X 2eme case: we are going to compare between these two sensors of dimension of output to post the results requested we are going to expose the curves which represent the deferentes temperatures (entry, exit, ambient) according to time in the two sensors, and then we will represent the curves of comparison in the same graph to clearly note the best output.

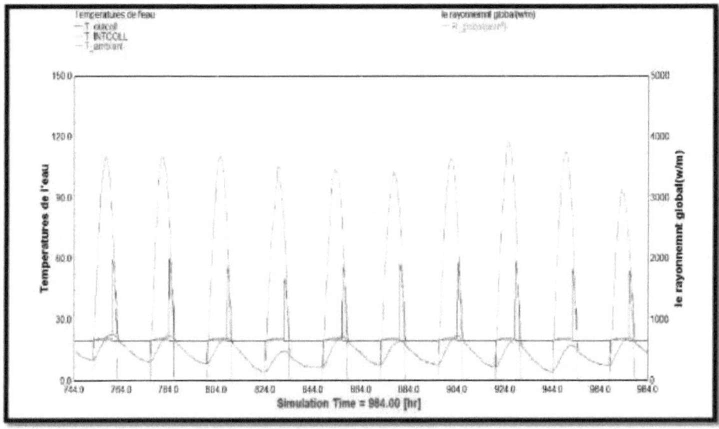

Figure (IV.7) shows the evolution of the 3 outlet and inlet temperatures and the ambient and global irradiation over time in the case of a glazed flat plate solar collector.

Figure (IV.7): shows the parameters influencing the collector's performance, based on the irradiation rate and temperature variation, not forgetting constant factors such as surface area, flow rate, etc.

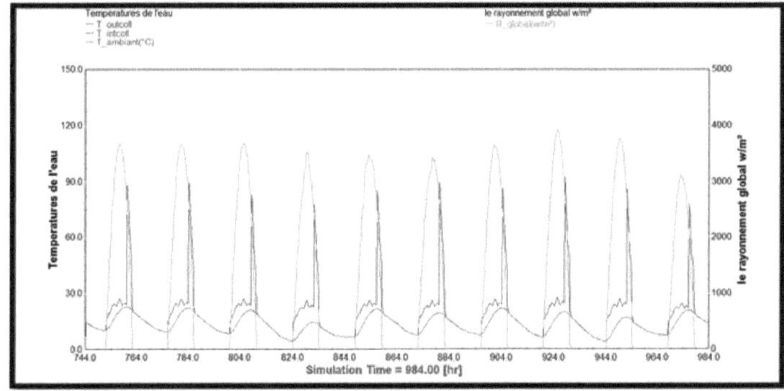

Figure (IV.8) shows the evolution of the 3 outlet and inlet temperatures and the ambient and global irradiation over time in the case of a flat-plate evacuated tube solar collector.

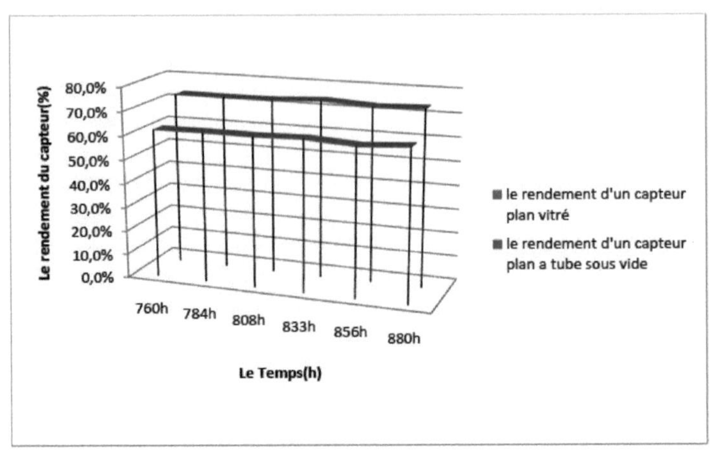
Le rendement du capteur(%)
80,0%
70,0%
60,0%
50,0%
40,0%
30,0%
20,0%
10,0%
0,0%
760h
784h
808h
833h
856h
880h
Le Temps(h)
le rendement d'un capteur plan vitré
le rendement d'un capteur plan a tube sous vide

Figure (IV.9) shows the two efficiency curves for the Type 1b and Type 71 sensors as a function of time (h).

IV.8 Analysis and interpretation:

On the whole, we note that the efficiency of a flat-glass collector is low, but above all deteriorates as the temperature rises. To improve the collector's efficiency, the reflective power of the glass surfaces should be reduced by using selective glass, reduce convection around the absorber by using an evacuated collector in the case of high temperatures, and reduce the temperature of the heat-transfer fluid by calculating the installation correctly, and then We note that the efficiency of the evacuated collector is much better than that of the flat-glass collector, and that the efficiency does not fall as quickly as the temperature rises. This is one of the reasons why evacuated collectors are used for high temperatures. Tube collectors have very small but highly protected inlet surfaces (the absorbers inside the tubes), which are very well insulated by the evacuated tubes, whereas flat-plate collectors have a large absorber inlet surface but are insulated only by the bottom and sides.

IV.9 The optimum inclination for this type of sensor (vacuum tube sensor).

We assume that the optimum orientation is due south azimuth α=0°C, and we will vary the angle of inclination of the collector to obtain the optimum inclination.

In this section we will look for the definitive optimal angle of inclination for one year for the production of hot water during the months of February and June, then winter and summer, based on this simulation of 3 angles (latitude, latitude-10°C, latitude+10°C in each period).

For β=27.87°C.

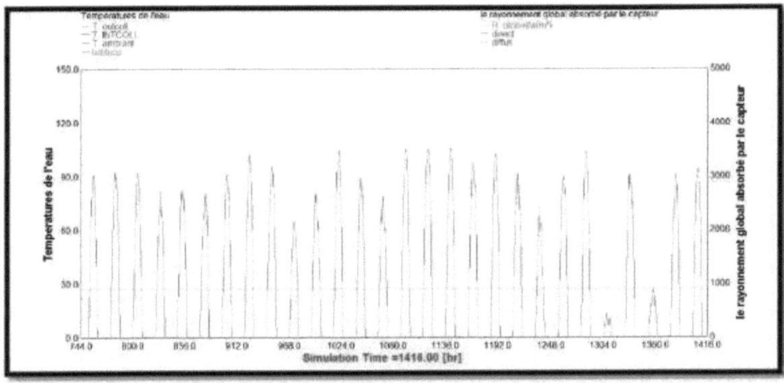

Figure (IV.10): Global radiation absorbed by the sensor as a function of time (month of February).

For β=37.87°C.

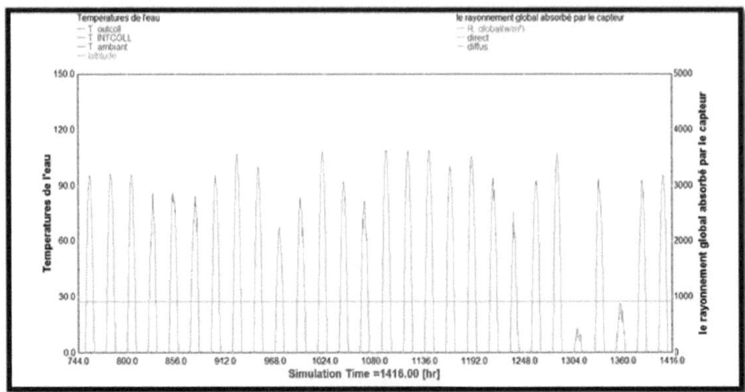

Figure (IV.11): Global radiation absorbed by the sensor as a function of time (month of February).

For β=17.87°C

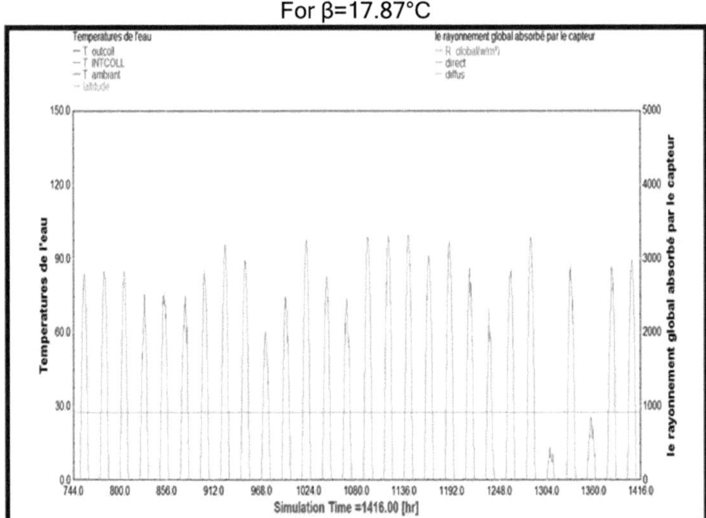

Figure (IV.12): Global radiation absorbed by the sensor as a function of time (month of February).

@ For β=27.87°C.

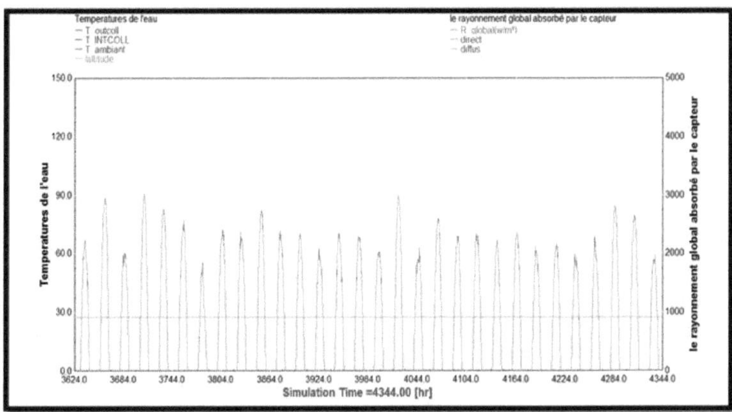

Figure (IV.13): Global radiation absorbed by the sensor as a function of time (the month of June).

@ For β=37.87°C.

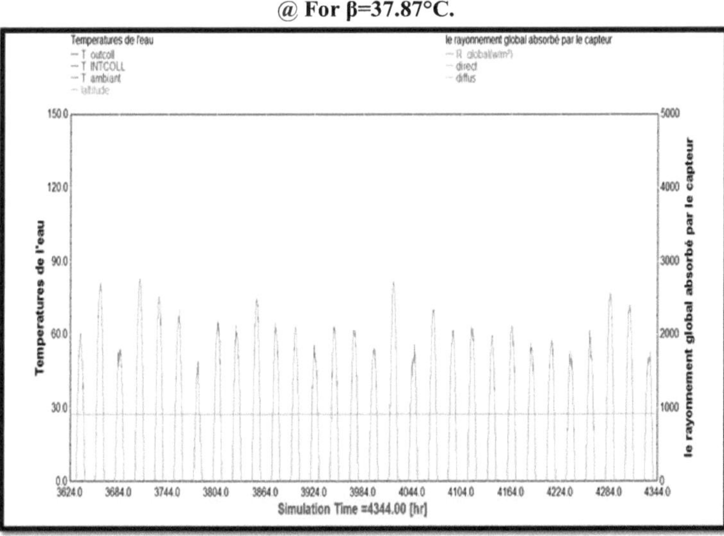

Figure (IV.14): Global radiation absorbed by the sensor as a function of time (month of June).

For β=17.87°C.

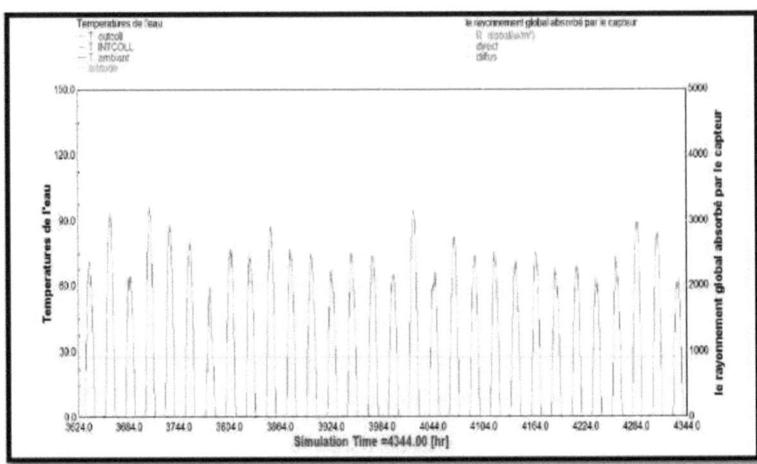

Figure (IV.15): Global radiation absorbed by the sensor as a function of time (the month of June).

Interpretation:

The aim is to **produce more heat in winter** and less heat in summer. So we're aiming for an optimum tilt in winter, as close as possible to 37.87°C. *The* best tilt is 27.87°C (local latitude). This tilt does not correspond to the maximum amount of sunshine, because with $2.7m^2$, production is surplus to requirements in summer and the panels are limited by requirements, so they do not produce at their maximum. As a result, it is preferable to increase production in winter and reduce it slightly in summer, to obtain maximum production over the year.

General conclusion:

The aim of this modest work is to optimise the yield of a solar water heater by using another type of solar collector, instead of installing a flat-plate collector, we will replace it with a vacuum tube collector to increase the yield of the installation, and the optimum angle during the year because the solar thermal collector is a device and a very important element in solar installations and finally we will be able to write the following results:

Superior efficiency: 70% of radiation absorbed. There is virtually no heat loss thanks to the vacuum in the tubes. Flat-plate collectors have the same absorption capacity, but the losses are much greater.

X **Solar absorption is not affected by the sun's angle of attack** because the tubes are round. Evacuated collectors are therefore effective from morning to night. Flat-plate collectors are very effective when the sun is in line with the glazing, but only for a few hours a day.

A reflector can be installed to increase the absorption surface.

."> **The ability to produce very high temperatures very quickly** (80°C and more), whereas flat-plate collectors are limited to 50-60°C (see graph). This has several implications: Lower storage volume. Possibility of heating a house with conventional radiators that operate at high temperatures.

The performance of vacuum collectors is not dependent on **wind or outside temperature**.

X The optimum angle during the year for a solar water heater is closer to or equal to the latitude of the location in the case of adrar is **β**= 27.87°C.

References

[1] **J.M Chassériau,** Conversion thermique du rayonnement solaire ; Dunod, 1984.

R. Bernard; G. Menguy; M. Schwartz, Le rayonnement solaire conversion thermique et applications ; Technique et documentation Lavoisier, 2 ème édition 1980.

[2] **S. Saadi,** Effect of operational parameters on the performance of a flat plate solar collector,

Magister thesis in physics; UMC, 2010.

[3] **M. Capderou,** Atlas solaire de l'Algérie, Tome 1, Vol. 1 and 2; OPU, 1987.

[4] **J.A Duffie and W.A Beckman**, Solar Energy Thermal Processes; 2nd edition, Wiley Inter science, New York, 1974.

[5] **P. Rivet**, Le Rayonnement solaire; CNRS.

[6] **A. Mefti; M.Y Bouroubi; H. Mimouni**, Evaluation du potentiel énergétique solaire, Bulletin des Energies Renouvelables, N° 2, P12, December 2002.

[7] http://www.inessolaire.com

[8] J, Bernard. Solar energy calculations and optimisation, Ellipse Edition Marketing (2004).

[9] A, Sfeir ; G, Guarracino. Ingénierie des systèmes solaires, Technique et Documentation, Paris (1981).

[10] A, Mefti, M, Y, bouroubi; H, Mimouni. Evaluation du potentiel énergétique solaire, Bulletin des Energies Renouvelables, N°2, p 12, décembre. (2002).

[11] F.bouhired .development of solar water heaters in algeria.(2002)

[12]: internet : www.tecsol.frSami, S, Lafri, D and Hamid, A., "Etude du comportement thermique d'une installation de chauffage d'eau collective", Revue des Energies. Renouvelables, Numéro spécial- Energies Renouvelables- Valorisation- pp 255-260 Tlemcen,(1999).

[13] Sami, S., "Etude et réalisation d'une installation de chauffage d'eau sanitaire d'une capacité de 1500 litres", Rapport interne -CDER- Décembre (2000).

[14] http://www.cder.dz/bulletin.

[15] Mr S.BEKKOUCHE. Modelling the Thermal Behaviour of Some Solar Devices. Option " Electronique et Modélisation " . Doctoral thesis. Abou-bakr- Belkaid University - Tlemcen *(2008).*

[16]: Y. JANNOT "***Thermique solaire***" October 2003.

[17] S.V. Joshi, R.S. Bokil, J.K. Nayak, "***Test standards for thermosyphon-type solar domestic hot, water system: review and experimental evaluation***" , Solar Energy 78 (2005) 781-798

Printed by Books on Demand GmbH, Norderstedt / Germany